Lecture Notes in Computer Science 1385
Edited by G. Goos, J. Hartmanis and J. van Leeuwen

Springer
Berlin
Heidelberg
New York
Barcelona
Budapest
Hong Kong
London
Milan
Paris
Santa Clara
Singapore
Tokyo

Tiziana Margaria   Bernhard Steffen
Roland Rückert   Joachim Posegga (Eds.)

Services
and Visualization

Towards User-Friendly Design

ACoS'98, VISUAL'98, AIN'97
Selected Papers

Springer

Volume Editors

Tiziana Margaria
Universität Passau, Fakultät für Mathematik und Informatik
Innstr. 33, D-93040 Passau, Germany
E-mail: tiziana@fmi.uni-passau.de

Bernhard Steffen
Universität Dortmund, Lehrstuhl für Programmiersysteme
Fachbereich Informatik
D-44221 Dortmund, Germany
E-mail: steffen@informatik.uni-dortmund.de

Roland Rückert
Bertelsmann OMS
Carl-Bertelsmann-Straße 161 O, D-33311 Gütersloh, Germany
E-mail: Roland.Rueckert@bertelsmann.de

Joachim Posegga
Deutsche Telekom AG, Technologiezentrum FZ122h
Am Kavalleriesand 3, D-64295 Darmstadt, Germany
E-mail: posegga@tzd.telekom.de

Cataloging-in-Publication data applied for

Die Deutsche Bibliothek - CIP-Einheitsaufnahme

Services and visualization : towards user-friendly design ; selected
papers / ACOS '98 ... [ETAPS '98]. Tiziana Margaria ... (ed.). -
Berlin ; Heidelberg ; New York ; Barcelona ; Budapest ; Hong Kong
; London ; Milan ; Paris ; Santa Clara ; Singapore ; Tokyo : Springer,
1998
  (Lecture notes in computer science ; Vol. 1385)
  ISBN 3-540-64367-2

CR Subject Classification (1991): C.2, D.2, H.5, H.3.5

ISSN 0302-9743
ISBN 3-540-64367-2 Springer-Verlag Berlin Heidelberg New York

© Springer-Verlag Berlin Heidelberg 1998
Printed in Germany

Typesetting: Camera-ready by author
SPIN 10632053   06/3142 – 5 4 3 2 1 0   Printed on acid-free paper

Foreword

The European conference situation in the general area of software science has long been considered unsatisfactory. A fairly large number of small and medium-sized conferences and workshops take place on an irregular basis, competing for high-quality contributions and for enough attendees to make them financially viable. Discussions aiming at a consolidation have been underway since at least 1992, with concrete planning beginning in summer 1994 and culminating in a public meeting at TAPSOFT'95 in Aarhus.

On the basis of a broad consensus, it was decided to establish a single annual federated spring conference in the slot that was then occupied by TAPSOFT and CAAP/ESOP/CC, comprising a number of existing and new conferences and covering a spectrum from theory to practice. ETAPS'98, the first instance of the European Joint Conferences on Theory and Practice of Software, is taking place this year in Lisbon. It comprises five conferences (FoSSaCS, FASE, ESOP, CC, TACAS), four workshops (ACoS, VISUAL, WADT, CMCS), seven invited lectures, and nine tutorials.

The events that comprise ETAPS address various aspects of the system development process, including specification, design, implementation, analysis and improvement. The languages, methodologies and tools which support these activities are all well within its scope. Different blends of theory and practice are represented, with an inclination towards theory with a practical motivation on one hand and soundly-based practice on the other. Many of the issues involved in software design apply to systems in general, including hardware systems, and the emphasis on software is not intended to be exclusive.

ETAPS is a natural development from its predecessors. It is a loose confederation in which each event retains its own identity, with a separate programme committee and independent proceedings. Its format is open-ended, allowing it to grow and evolve as time goes by. Contributed talks and system demonstrations are in synchronized parallel sessions, with invited lectures in plenary sessions. Two of the invited lectures are reserved for "unifying" talks on topics of interest to the whole range of ETAPS attendees. The aim of cramming all this activity into a single one-week meeting is to create a strong magnet for academic and industrial researchers working on topics within its scope, giving them the opportunity to learn about research in related areas, and thereby to foster new and existing links between work in areas that have hitherto been addressed in separate meetings.

ETAPS'98 has been superbly organized by José Luis Fiadeiro and his team at the Department of Informatics of the University of Lisbon. The ETAPS steering committee has put considerable energy into planning for ETAPS'98 and its successors. Its current membership is:

André Arnold (Bordeaux), Egidio Astesiano (Genova), Jan Bergstra (Amsterdam), Ed Brinksma (Enschede), Rance Cleaveland (Raleigh), Pierpaolo Degano (Pisa), Hartmut Ehrig (Berlin), José Fiadeiro (Lisbon), Jean-Pierre Finance (Nancy), Marie-Claude Gaudel (Paris), Tibor

Gyimothy (Szeged), Chris Hankin (London), Stefan Jähnichen (Berlin), Uwe Kastens (Paderborn), Paul Klint (Amsterdam), Kai Koskimies (Tampere), Tom Maibaum (London), Hanne Riis Nielson (Aarhus), Fernando Orejas (Barcelona), Don Sannella (Edinburgh, chair), Bernhard Steffen (Dortmund), Doaitse Swierstra (Utrecht), Wolfgang Thomas (Kiel)

Other people were influential in the early stages of planning, including Peter Mosses (Aarhus) and Reinhard Wilhelm (Saarbrücken). ETAPS'98 has received generous sponsorship from:

Portugal Telecom
TAP Air Portugal
the Luso-American Development Foundation
the British Council
the EU programme "Training and Mobility of Researchers"
the University of Lisbon
the European Association for Theoretical Computer Science
the European Association for Programming Languages and Systems
the Gulbenkian Foundation

I would like to express my sincere gratitude to all of these people and organizations, and to José in particular, as well as to Springer-Verlag for agreeing to publish the ETAPS proceedings.

Edinburgh, January 1998   Donald Sannella
                                 ETAPS Steering Committee chairman

Preface

This volume contains selected papers from three international workshops:

- *ACoS'98, International Workshop on Advanced Communication Services*, which took place on April 3–4, 1998,
- *VISUAL'98, International Workshop on Visualization Issues for Formal Methods*, which took place on April 4, 1998, and
- *AIN'97, 2nd International Workshop on Advanced Intelligent Networks*, which took place on July 5–6, 1997,

which all focus on specific techniques for *user friendly design*.

ACOS'98 and VISUAL'98 were associated with TACAS'98 (*International Conference on Tools and Algorithms for the Construction and Analysis of Systems*), one of the main conferences of ETAPS'98, the first *European Joint Conferences on Theory and Practice of Software*, which took place at the Gulbenkian Foundation in Lisbon, Portugal, March 28 – April 4, 1998.

AIN'97 took place in Cesena, Italy, and was associated with last year's ICALP, *International Colloquium on Automata, Languages and Programming*, July 7–11 in Bologna, Italy.

The common goal of all three workshops is to support technology transfer, and this in three ways:

- *directly*, as each of them defines a field of application for formal techniques,
- *indirectly*, by focusing on applications which themselves aim at simplifying the use of complex techniques, thus making them acceptable and available to a wider public, but also
- *on the metalevel*, by addressing techniques which themselves are applicable to improve the automatic support for technology transfer in the respective application fields (bootstrapping effect).

In particular, the scope of these workshops covers both complex designs and algorithms to offer technical support towards a wider use of complex systems (typically dealt with at AIN and ACoS), as well as techniques for simplifying the use of systems by adequate visualization (VISUAL). The potential for a fruitful dialogue among participants of both communities was evidenced by the fact that ACoS and VISUAL shared common invited speakers and featured a joint session. It is our pleasure to remark that our hopes became reality: reflecting the balanced composition of the program committees, contributions and participation came equally significantly from industry and academia, thus fostering a true exchange and discussion of ideas.

The volume contains contributions by Maurizio Decina and Pamela Zave, invited speakers for AIN'97, and by Alain Lardenois, common invited speaker for ACoS'98/VISUAL'98, together with a selection from the regular papers presented at the workshops.

ACoS'98 and VISUAL'98 were hosted by the University of Lisbon, and, being part of ETAPS, they shared the excellent sponsoring for this overall event. AIN was hosted by the University of Bologna at Cesena, and sponsored by Siemens-Nixdorf Informationssysteme AG and Cassa di Risparmio di Cesena.

Finally, warm thanks are due to the program committees and to all the referees for their assistance in selecting among the contributions, to Donald Sannella for mastering the coordination of the whole ETAPS, to José Luiz Fiadeiro, Roberto Gorrieri, and their teams for their brilliant local organizations, and, last but not least, to Claudia Herbers for her professional assistance during the last months and for her first class support in the preparation of this volume.

March 1998

Tiziana Margaria
Bernhard Steffen
Roland Rückert
Joachim Posegga

Program Committees

ACoS'98

Sahin Albayrak (D)
Friedrich-Karl Bruhns (D)
Maurizio Decina (I)
Reinhard Gotzhein (D)
Nancy Griffeth (USA)
Jens Gutsche (D)
Bengt Jonsson (S)

Tiziana Margaria (D)
Walter Ozinger (A)
Manfred Reitenspieß (D)
Marco Rocetti (I)
Roland Rückert (D, Co-Chair)
Bernhard Steffen (D, Co-Chair)
Pamela Zave (USA)

VISUAL'98

Lou Feijs (NL)
Kathi Fisler (USA)
Tiziana Margaria (D, Co-Chair)
Louise Moser (USA)

Doron Peled (USA)
Joachim Posegga (D, Co-Chair)
Peter Reintjes (USA)
Dave Robertson (UK)

AIN'97

Jane E. Cameron (USA)
Reinhard Gotzhein (D)
Nancy Griffeth (USA)
Bengt Jonsson (S)
Tiziana Margaria (D)

Manfred Reitenspieß(D)
Marco Roccetti (I)
Luigi Santagostino (I)
Bernhard Steffen (D, Chair)
Giorgio Tenerini (I)

Referees

Maria Elena Bonfigli (AIN'97)
Volker Braun (AIN'97)
Jan Bredereke (ACoS'98)
Carla Capellmann (VISUAL'98)
Stelvio Cimato (AIN'97)
Renzo Davoli (AIN'97)
Mauro Gaspari (AIN'97)
Shriram Krishnamurthi (VISUAL'98)
Cosimo Laneve (AIN'97)

Gustaf Naeser (ACoS'98)
Jan Nyström (ACoS'98)
Fabio Panieri (AIN'97)
Paola Salomoni (AIN'97)
Jochen Schiller (AIN'97)
Torge Schmidt (ACoS'98)
P.H. Schmitt (VISUAL'98)
Vincenzo Scinicariello (AIN'97)
Harald Vogt (VISUAL'98)

Table of Contents

Invited Lectures

ACoS'98 & VISUAL'98

AIN'97

Selected Papers

ACoS'98

VISUAL'98

AIN'97

The Web Impact: A White Paper

Alain Lardenois
PLAISIR, France
lardenois.pla@sni.de

Paradigm Shift due to Web Evolution

The success of the World Wide Web (WWW) is since some time creating an important redefinition of strategies in the computer industry. It can well be compared to the paradigm shift of mainframe oriented computing to decentralized computing. The generated trends will affect most of our current models and architecture:

1. The trend to revive a kind of mainframe architecture affects heavily the client/server model and develops the needs of very fast Intranet/INTERNET networks.
2. The inter-operability between applications is standardized with universal HTML data formats and portable JAVA applications.
3. A renewal of the client concept can be noticed (Network computer, settop boxes, cash devices, smart cards, telephones, ...).
4. User interface standards are changed, from windows or CDE to Mosaic/JAVA.
5. Changes are visible in the software distribution concept as well as in the sales model of most of the companies including the computer industry.
6. The need for new types of applications is generated.
7. Services are delocalised and decentralised.

Network Capabilities will Drive Evolution

The changes will not be immediate, because the natural inertia of customers, and the speed of evolution is heavily linked to the evolution of the network capabilities. The first boom is expected on the network side: The network capabilities both for Intranet and INTERNET will increase dramatically over the next years. The increase of the available bandwidth will naturally move all traffic including voice and later videos on this new data highways. As far as computers are concerned, the integration of data switches or phone switches as part of a company's « groupweb » concept will turn an opportunity into a real added value. The existing applications will first stay in their computing environment, but will all avail of gateways to the WWW which thereby will become the standard user interface method.

Trend to Large Mission Critical Servers

The size of the computer will tend back to large mission critical servers, first in the Intranet environment, and then later for public applications. Typically the Web is at the same time a centralization of information as well as a delocalisation of it, thanks to a fast network based on ISDN, ATM, LAN, and other high speed interconnects (SCI). Systems will be part of a company cluster environment. The security of access to critical applications is ensured by internal firewalls and crypting techniques. High speed interconnects such as SCI or other clustering techniques will be used to implement company backbones, to which all the dedicated application servers are connected. A new architecture concept will appear: based on the high connectivity with high speed interconnects (SCI / ATM) switches will be used to connect CPU/memories supporting applications, to I/O (WAN,LAN, ISDN, ...), and intelligent application engines like databases, archive, search, ...

Move to JAVA as New Application Programming Environment

Some new dedicated servers might appear, based on proprietary optimized packages like database engines, backup engines, printing engines or archive engines. This new model may give an additional kick to UNIX being a basis for new efficient operating systems based on MICROKERNELs and being used in packaged engines. The Microsoft client/server model will change progressively, and Microsoft has changed dramatically its organization to face this new unexpected challenge. The main evolution is the move towards JAVA as the new application programming environment thereby replacing native Windows applications. In addition, JAVA is a concept where the generated software is independent of the platform and can be compiled and being put into operation on the fly.

As far as databases are concerned, the current model of relational databases, which are based on predefined data structures will tend to move to databases able to handle unformated and fluctuating data structures such as object oriented databases or search engines used for query and retrieve.

Intelligent Agents Allow Personalisation of Data and Services

The concept of intelligent agents, running both on the servers and on the client and being able to inter-operate, will allow users to create a personalized agent. Such an agent will automatically collect, sort, prepare the data the user is interested in. The Web with its huge amount of unformated information will be used as a database. A banking agent will be discussed as an example:

On the bank server the system will use a customizable clone, that the customer calls once to provide his personal information such as stocks the user is interested in. The bank will update the clone with the amount of stock of each type this customer owns. The server also takes advantage of an agent clone, which is able to extract automatically information from any stock exchange server in HTML format. At any time the customer can download the personal agent (updated on the bank server each day) and locally execute it for getting dynamic statistics, autogenerated graphics, simulation results etc.

The Web will Invade Business Applications

Besides standard databases and some legacy applications, the Web will invade all types of business applications from ordering, logistics, cash handling, statistics, dataware housing, to customer information desks, hot lines, try and buy, demos, etc. Because the data model between the mentioned business application is very similar, the data can be separated from the applications. This means that the change from one application to a better or less expensive one will be much faster and less painful for a customer and will not change the user interface. Therefore the natural inertia will decrease heavily creating many new opportunities and challenges in the business of manufacturers and independent software vendores (ISVs).

Slim Technologies Driven by Web

Multiple gateways to existing applications will rapidly appear on the Web. Therefore the clients, even if they remain first on standard PCs, will tend to become slimmer. The consequence is to push PCs down to smaller configurations with less memory and disk space. This smaller demand of memory and disc may push the prices of these devices down, and slow down the demand for fast desktop PCs. As far as new clients are concerned, new terminals are under development like

- WEB GSM telephones,
- new slim terminals,
- settop boxes eventually,
- or home banking devices.

The PC will continue to develop in the laptop area, mainly because the telephone bandwidth is still not sufficient to support slim clients in the network. This pushed business of batteries and low power technologies. The phones (fixed and portable) will appear with built-in light Web access (Handheld Device Markup Language, HDML), pushing up the trend to the integration of voice and data.

Web Drives Multimedia Use

The desktop concept (CDE or Windows) will become HTML compliant, including systems and network administration. In addition, the availability of Java applets will create new user interfaces with 3D graphic presentation and intuitive access widgets. Also HTML compliant office packages are rapidly showing up. E-mail and archive capabilities will be managed by WWW servers. User defined applications (Spreadsheets, publishing, ...) will be able to import unformated text from any HTML page. As far as E-mail is concerned, multimedia E-mail including voice mails and video mails are evolving. As far as vocal messages invade the computer mail area, the demand for voice recognition as well as text-to-speech techniques will rapidly increase as a standard feature inside Web servers as well as clients (e.g. mobile phones).

The telephone will be interfaced straight to the Web by gateways. This results in the availability of unified E-mails including such diverse media as voice, fax, data, video. By registration via a browser, a customer will be able to interact with the E-mail box using the telephone, but also standard Web browsers. Moreover, these techniques will allow the development and rapid installation of Web integrated call centers (interaction centers), where the agent station is a multimedia PC or NC, and the client can be either on the phone or on a browser. Demands emerge in the Intranet business to handle all the company communications including: E-mails, fax, office communication, archiving, search engines, software distribution, telephony interfaces and eventually video on demand. From the administration perspective, telephony gateways and switches, LAN routers, gateways to existing applications, firewalls and automatic encryption/compression to protect internal data need to be addressed. This system price is expected to cost a similar price as a telephone interface i.e. in the range of 500$/2000$ per user (based on user communities of more than 1000 users. Small to medium enterprises will outsource these services to service providers to take advantage of the developing technology at affordable costs. In the Intranet environment, the client computer not necessarily needs to be „fat", because the base network (typically LAN) can support the bandwidth required by the above mentioned applications.

E-Business Forces Cross-National Awareness

The tendency to use application-on-demand services through the network will enforce the use of electronic cash. Electronic commerce will first be applied to software accounting because of its inherent nature. The natural linkages between customers and suppliers will be service providers, which will offer the network and/or services. The current model of suppliers, distributors, value added resellers can therefore change dramatically. The actual model with four layers in the distribution chain may be compressed down to two with strong impacts on prices and margins. The marketing of software products will also dramatically change with on-line marketing, demonstrators on the Web or try-and-buy approaches easily accessible. In addition, there is an new opportunity for European software companies (independent software vendors, ISVs) due to the localization requirements and multi-national environment. It is highly likely that search engines will become the focal point of advertisement starting with applications over the net. *It therefore will become very important to be represented in a European search engine to advertise company products.*

Intranet Business and Development Extends into Internet

New applications are appearing on the market (as it was the case when the minitel appeared in France some 15 years ago). The very first were search engines, offering catalogues, business information, weather reports, stock exchange data, search by name capabilities on the network. Very fast, reservation engines for flights, hotels, E-mail boxes, discussion forums, and electronic commerce followed up. Pricing and user habits, however, are decisive factors for the success of such new developments.

These foreseeable developments are leading to a decentralization of the distribution of goods, therefore increasing the need for multi-media mails, video conferencing, and shared applications extending the developing intranet business into the Internet. Nevertheless, the transition from traditional markets and buying habits into Internet based business will not start growing on a large scale before 1999. Major limiting factors are unavailable network capabilities, costs of accessing the networks, but also the associated legal regulations and user habits.

Competition will Increase

All good, but the size of the market is raising immense competition, as most of the current players will need to adapt to survive in the mature stage of technology. Also, the trend to Internet could also foster new, currently unknown architectures and technologies. The criteria of the technology winners could be sorted as follows:

1. Who is innovative and can maintain the pace of innovation?
2. Who is able to support adversity during the (potentially long) transition period towards the INTERNET?
3. Who will be able to be a global player? The US, Europe, and Asia Pacific markets are expected to be one third each.
4. Who is able to follow their customers in the migration path? Gateways will be required to bridge the gap between supporting legacy applications in parallel to the development of new ones. Still, a continuous operational quality must be maintained.
5. Who is able to integrate within the Web a global communication system such as providing voice, fax, video, data services and support for financial services with high security requirements.

The following major players can be foreseen in Europe.

- Cable providers, TV broadcast providers
- Electronic commerce providers
- Financial services
- Content providers
- Application software suppliers
- Security
- Network and data equipment providers
- Integrators

Opportunities and Challenges at the Same Time

As has been explained above, tremendous growth and change can be expected not only in the commercial environment due to the rapid introduction and growth of Web technologies. Also the social environment is expected to change. E.g. the trend to remote work places will have an impact on the way people interact with each other in a business environment. The agent technology pushes strongly to enter the Web and provide new, important services in the Web distributed environment. It can play a major role to make the Web more useful and to take advantage of the forseeable global village. Its impact on the social environment still needs more research and analysis, which in turn may lead to new, currently not foreseen technical development and change.

Literature

[1] White Paper: Open Service Node: A Service Control System for Intelligent Networks; Intellitel Communications Ltd., Lappeenranta, Finland; 1997; http://www.intellitel.com/osnwhite/osnwhite.html

[2] Ursula Hinkel, Wolfgang Kellerer, Peter Sties: Multimedia Anwendungen in der Telekommunikation - Szenarien und Abläufe -; Technical Report TUM-LKN-TR-9702; Munich, D, Sept. 1997

[3] Manfred Reitenspiess: Web, Multimedia und Telekommunikation: Ein Plus an Service und Funktionalitaet; Siemens Nixdorf Informationssysteme AG, Munich, D; 1997; http://smartcom.mch.sni.de

[4] Maurizio Decina, Vittorio Trecordi: Convergence of telecommunications and computing on networking models for integrated services and applications; Proceedings of the IEEE, Special Issue, 1997

[5] Grünbuch zur Konvergenz der Branchen Telekommunikation, Medien und Informationstechnologie und Ihren Ordnungspolitischewn Auswirkungen; European Commission; Brussels B, Dec. 1997; http://www.ispo.cec.be/convergencegp

[6] Voice of the LAN: The Distant Ring of LAN Telephony; Data Communications White Paper, Vol. 12, No. 3, April 1997

[7] Agents on the Web - Catalyst for E-commerce; Ovum Reports, Christine Guilfoyle, Judith Jeffcoate, Heather Start (Edts.); Ovum Ltd., London UK, 1997; ISBN: 1 898972 12 5

[8] John Davison, Jolanda Goverts: Network Computing - Opportunities for the Consumer Market; Ovum Reports, Ovum Ltd., London UK, 1997; ISBN: 1 898972 72 9

'Calls Considered Harmful' and Other Observations: A Tutorial on Telephony

Pamela Zave

AT&T Laboratories—Research

pamela@research.att.com

Abstract. The software application domain of customer-oriented telephony is worth the attention of specialists in formal methods. Primarily this paper is a tutorial about customer-oriented telephony. It also includes observations about how this domain should be formalized, including critiques of some popular approaches.

1. Introduction

In the evolution of formal methods for software engineering, the time has come to develop formal methods for particular application domains [10,24]. This paper concerns the domain of customer-oriented telephony, i.e., software producing the externally observable behavior of voice telecommunications systems.

Customer-oriented telephony is worth the attention of industrial researchers and academicians alike, for the following three reasons:

Accessibility. Everybody uses telephones. Many characteristics of telephone systems are determined by international standards. Many consumer products now offer significant telephony functions. Anyone can study this domain, even without access to private intellectual property.

Importance. Telecommunications is widely predicted to be one of the key industries of the 21st Century. Both its political/economic context and its technological base are changing rapidly.

Trouble. Customer-oriented telephony software has all the usual problems of complex, long-lived, distributed, high-performance software systems. Several characteristic problems, such as feature interaction and the intertwining of separable concerns, are particularly visible and severe.

Customer-oriented telephony has already been the subject of much research activity, including invention of new specification languages and methods, implementation of new tools and environments, workshops, case studies, and other research projects. Nevertheless, there seems to be widespread ignorance of some of the basic principles and technological foundations of telephony. Naïveté mars much of the published work, and either compromises its usefulness or makes its usefulness difficult to evaluate.

This paper is a tutorial on customer-oriented telephony for the formal-methods community. It is intended to give someone who is interested in this software domain a good start, with enough depth and perspective to avoid egregious errors. Sections 2 through 5 delimit the domain and explain the relevant facts. Section 6 presents observations and conclusions about formal descriptions of this domain, including critiques of some popular approaches. It explains why the *call model,* which is the foundation of most formal descriptions of telephony, is limited and potentially harmful.

Although there is plenty of tutorial material to be found in the networking literature, the presentation of telephony in this paper is unique. For one thing, its audience is different from the expected readership of a networking journal. For another thing, there is an unusual emphasis on relating different kinds of telephone system: how they are similar, how they are different, and how they interact. This emphasis is necessary to convey the crucial information in a small space, but it also has the important advantage of revealing the inherent coherence of the application domain. As a result of the coherence newly revealed here, it seems possible that all kinds of telephone system can be specified with the same techniques.

2. Boundaries of the Domain

Figure 1 shows the world-wide telephone network decomposed into two (highly distributed) machines, one performing voice transmission and one performing customer-oriented telephony. We are interested in the behavior of the upper machine, which is implemented exclusively in software.

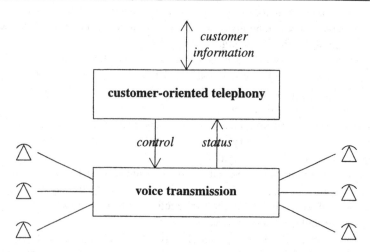

Figure 1. A decomposition of the world-wide telephone network.

The relationship between the two machines is similar to the relationship between two adjacent layers in the OSI Reference Model.[1] The customer-oriented machine sends control commands to the voice-transmission machine. The voice-transmission machine sends status information to the customer-oriented machine, including command results and notifications of stimuli from telephones. The *call-*

---

[1]The split between the two machines shown in Figure 1 appears to fall within the application layer of the OSI Reference Model [19].

processing function of the customer-oriented machine is to emit control commands at this interface.

The customer-oriented machine also has other interfaces and functions, all concerning customer information. It must accept information about customer identity, service preferences, payments, etc. It must emit information about bills, service usage patterns, etc.

There is plenty of software in the voice-transmission machine, so the decomposition shown in Figure 1 does not separate software from hardware. Rather, it separates software functions directly observable by customers from software functions for resource management: hardware monitoring, hardware fault diagnosis, resource allocation (including network routing and the creation of voice paths), performance tuning, and interaction with operators who participate in the resource-management functions.

Telecommunications is concerned with transmission of data and multimedia as well as plain voice (telephony); often all of these media are transmitted on the same physical network. Why separate telephony from the rest of telecommunications?

From the perspective of users, data transmission is fundamentally different from telephony. Multimedia communication, on the other hand, is an extension of telephony. I recommend attacking the problems of telephony first because they are extremely difficult in themselves, yet have a certain familiarity and coherence that may enhance intuition. Better to solve these first and then extend the results to multimedia (which appears possible [18]) than to attempt everything at once and get nowhere.

3. Overview of Voice Transmission

This section is an overview of the voice-transmission machine. The information in it is necessary for two reasons: (1) the call-processing function of the customer-oriented telephony machine consists of controlling the voice-transmission machine, and (2) the capabilities of the voice-transmission machine limit the services that the customer-oriented machine can offer to its customers. The voice-transmission machine is also the least familiar part of telephony. The customer-information interface, in contrast, can easily be imagined by any computer scientist.

3.1. Components

Figure 2 shows some of the components of the world-wide voice-transmission machine (or "network," since it is highly distributed).

A *telephony device* is an input/output device for voice.[2] It might be a conventional telephone, speakerphone, cordless telephone, mobile telephone, personal computer equipped with a speaker and microphone, fax machine, answering machine, or many other things. The key characteristic of a telephony device is that it supports a single two-way voice channel. This makes sense because sound usually

---

[2]There is no industry-standard term for a telephony device. *Customer-premises equipment* comes close in meaning, but also includes PBXs (see Section 4).

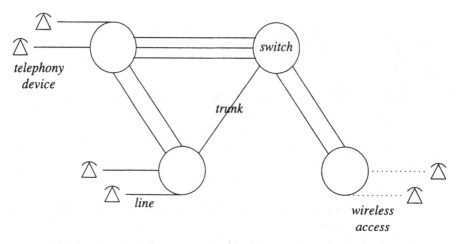

Figure 2. Some components of the world-wide voice-transmission network.

passes between the device and a human head through the air, and there would be no way to maintain the acoustical isolation of two voice channels.

A telephony device is usually connected to the network by a *line*, dedicated to the device and also capable of supporting a single two-way voice channel. The exception is a mobile telephone, which has wireless access. It is connected to the network (whenever it is connected) by a two-way radio channel allocated to the telephone only for the duration of the access episode.

When multiple (extension) telephony devices are plugged into the same line, the rest of the network perceives their collective actions as the actions of one device. Since the telephone systems in the network cannot distinguish the presence or absence of extensions, specifications can ignore their existence completely.

A line leads from a telephony device to a *switch*, which is a voice-handling node of the network. Switches are connected by many *trunks*, each of which is like a line in supporting a single two-way voice channel. A switch is capable of connecting any two trunks or lines coming into the switch, for the purpose of creating a voice path (see Section 3.3).

The components shown in Figure 2 are the ones that support voice transmission directly. In other words, Figure 2 is a picture of a circuit-switched network. In addition to the circuit-switched network, there is a separate packet-switched *signaling network* used for transmitting control messages among the switches.

In addition to signaling among switches, there is also a need for signaling between telephony devices and switches. If the device's line is digital, then it has a two-way signaling channel separate from its voice channel. If the device's line is analog, then the voice channel is also used for signaling (see Section 3.2).

3.2. Voice Processing

Many switches contain (or are closely associated with) special hardware devices for voice processing. For example, a *three-way conference bridge* is a device that connects three two-way voice channels. It mixes its three voice inputs so that each of its three voice outputs is the normalized sum of the other two inputs. Conference bridges can be built to mix almost any number of voice channels. The important thing about conference bridges is that a more-than-two-way conversation is never possible without one.

Another example of voice processing would be recording and playback of speech. By far the most common use of voice processing, however, is for *in-band signaling*, in which control signals are transmitted on the voice channel. In-band signaling is defined in contrast to *out-of-band signaling*, in which control signals are transmitted through a separate signaling channel or network.[3]

In-band signals to a telephony device take the form of tones or announcements (a voice-processing device generates the tones and either plays recorded announcements or synthesizes speech from textual announcements). In-band signals from a telephony device usually take the form of touch-tones, which are versatile and easily detected by voice-processing hardware. Keywords and sound/silence transitions can also be recognized as control signals.

3.3. Voice Paths

A *voice path* allows persistent, two-way voice transmission between two endpoints.[4] An endpoint is usually a telephony device or a voice-processing device within a switch. An endpoint might also be just a loose end within a switch, in which case the other endpoint is *on hold*. A voice path can pass through any number of switches and trunks.

Many telephony features manipulate voice paths, as illustrated by the sequence from (a) to (d) in Figure 3. In (a) there is a voice path between the left and right telephones. In (b) the left telephone has put the right telephone on hold and obtained a path to a middle telephone; there are now two voice paths. In (c) the three telephones are conferenced, each having its own path to a conference bridge, so that there are now three voice paths. In (d) the left telephone has ordered a transfer, dropping out of the conversation while leaving the other two telephones joined by a single voice path.

Let us consider how a path such as the one shown in Figure 3(a) could be set up and torn down. The protocol used to bring each line or trunk into the path could be different, but all the protocols are similar. Figure 4(a) is a "message sequence chart" illustrating one kind of setup. The control signals between any two nodes refer to the line or trunk that is being added to the path, and travel on signaling channels associated with the line or trunk (which might be the same as the voice channel of the line or trunk, in the case of in-band signaling).

Figure 4(a) shows only the two most important types of control signal,

---

[3]Voice processing is usually called *signal processing*. I have used a different term to distinguish voice processing in general from its particular use for in-band signaling.

[4]A voice path is a *circuit* in the terminology of circuit-switched networks.

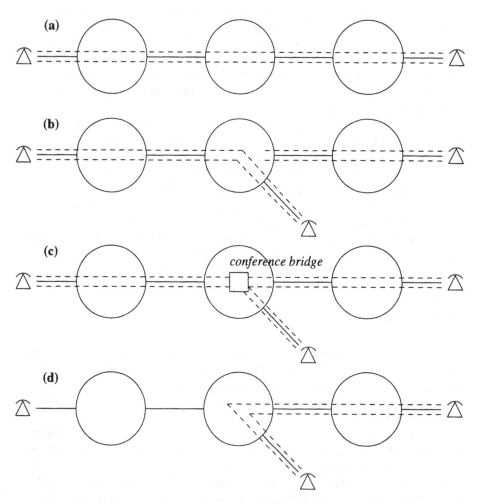

Figure 3. Some examples of voice paths.

generically named *request-path* and *complete-path*, although the protocol between any two particular nodes might use other signals as well. The path is set up when the last *complete-path* signal reaches the left telephone.

When each switch in Figure 4(a) receives a *request-path* signal, there are decisions to make. At the level of the voice-transmission machine, there is always a physical routing decision. At the level of the customer-oriented telephony machine, there may be a decision to redirect the path (perhaps because the current destination has requested call forwarding), or to treat the request specially in some other way.

The path can also be set up in stages, as shown in Figure 4(b). Here switch *S2* takes a very active role. It first completes setup of a path from the left telephone to

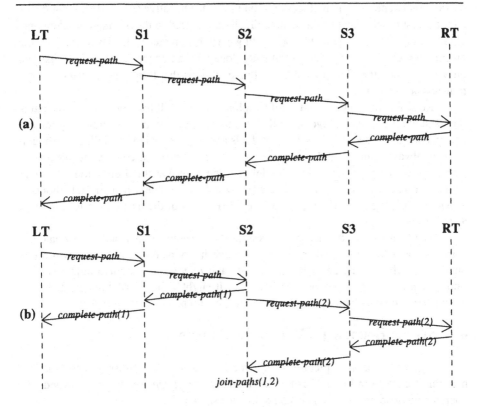

Figure 4. Two scenarios for setting up a voice path.

itself, then initiates setup of another path from itself to the right telephone. When the second path is complete it joins the two paths, creating a single voice path between the left and right telephones.

Why set up a voice path in multiple stages? There are many possible reasons, but the most common one is the need for in-band signaling. Suppose that, when the path request reaches *S2*, the customer-oriented machine needs more information about the request than comes with the request signal. Further suppose that the protocol or protocols active between *S2* and the left telephone have no built-in features for obtaining and transmitting the additional information. Then the only remaining option is to complete the voice path between *S2* and the telephone, so that *S2* can solicit and collect the additional information through in-band signaling. An example of this situation will be given in Section 4.

Many other protocol details can influence a specification of telephony services, depending on its level of abstraction. Consider, for example, how a busy tone is produced. In one common method, the switch directly connected to the destination (busy) telephone completes setup of the voice path and then connects its end of the path to its own busy-tone generator. The advantage of this method is that the busy tone heard at the originating telephone corresponds to the geographic region of the destination telephone. In another common method, a negative acknowledgment of the path request is sent back to the switch directly connected to the originating telephone. Upon receiving this signal, the originating switch completes the voice path to the originating telephone (if it has not already done so) and connects the path to its own busy-tone generator. The advantage of this method is that it minimizes use of voice-transmission resources.

Roughly speaking, providers of telephone service bill for completed voice paths (and sometimes feature usage as well). But there are many details and exceptions to be considered. In the two busy-tone methods above, there is never a bill, even though there is always some completed voice path. In the case of a completely successful path request, the path is usually completed at the time that the destination telephone starts to ring, to ensure that a voice path will be available when the telephone is answered. Billing does not usually begin, however, until the destination telephone is actually answered.

The teardown of a voice path is similar to its setup. Either end can initiate the teardown, and the other end must acknowledge it. A path can be torn down all at once, in which case the generic *release-path* and *release-ack* signals might have the same general pattern as shown in Figure 4(a). It might also be torn down in stages, in which case the signals might exhibit the same general pattern as Figure 4(b).

4. System Boundaries Within the Domain

Figure 5 is another picture of the world-wide voice-transmission network. Its main difference from Figure 2 is that the boundaries of *telephone systems*, owned and operated by different *service providers*, are also shown.

A *private branch exchange (PBX)* is a private switch, usually found on the premises of a business or institution. A *local exchange carrier (LEC)* provides local service; it may run a local network or simply a single switch. An *interexchange carrier (IEC)* provides a long-distance network. A *national system* provides telephone service for an entire country, combining the functions of LEC and IEC systems. A *cellular system* provides mobile service; it may reach the rest of the world through any other type of system.[5,6]

Needless to say, any particular software-development project is going to be confined within the boundaries of a single telephone system. We are now in a position to understand the similarities and differences among the various types of telephone system.

One difference is that most telephone systems are distributed, while some (e.g.,

---

[5]Other combinations are possible. For example, in the United States, there will soon be systems that act like national systems in the sense of combining the functions of LEC and IEC systems, and that act like IEC systems in the sense of having direct competition and needing the cooperation of other service providers for access to some local telephones.

[6]An Internet-based telephone system is harder to include in this diagram, both because the Internet is an overlay network and because it is a data network. When a user of an Internet telephone service wishes to speak to someone who does not subscribe to the same Internet telephone service, then his call goes to the public network through a *tail-end hop-off* feature. Tail-end hop-off is exactly like the interface between a LEC system and an IEC system.

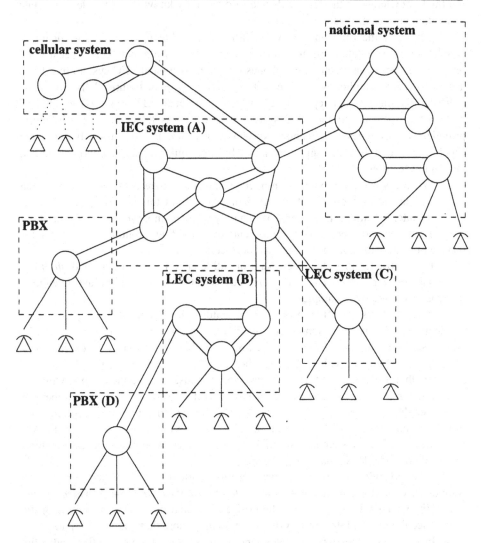

Figure 5. Some systems and components of the world-wide voice-transmission network.

PBXs) are not. This difference is discussed in Section 6.1.

The most important distinction among telephone systems is that some systems have direct access to telephony devices, while others do not. There is often a special, knowledge- and feature-rich relationship between a telephone system and a direct-access device. The system knows what kind of device it is, and whether it is busy or idle. If the device is complex (e.g., has many buttons) then the protocol used on the line to that device probably has similar complexity (many signals). In contrast, a system that interacts with a telephony device through another system usually has minimal knowledge of the remote device, and can only know about the device's state what another system tells it.

Although the difference between direct and indirect access is significant, it should not be overstated. *No* telephone system is free of indirect access, because no telephone system can reach all telephones by itself. A typical LEC system has direct access to many telephones, but the LEC system (e.g., *B* in Figure 5) has the same indirect relationship to telephones connected to a local PBX (e.g., *D*) as an IEC system (e.g., *A*) has to the LEC system's direct-access telephones. Furthermore, providers of all types of telephone system now aspire to offer roughly the same capabilities, including routing, screening, billing, multiplexing, conferencing, and messaging features.

When a voice path includes an intersystem trunk, there is cooperation between the two adjacent systems to set up the path. At present this is essentially the only means of interoperation. Interoperation can only be made more powerful by enriching the current protocols between systems, and by providing economic incentives for service providers to support the enriched protocols.

In the absence of richer interoperation, all telephony features beyond *plain old telephone service (POTS)* are implemented strictly within individual telephone systems. Here are two examples.

If an IEC system offers a credit-card feature, then setup of voice paths using that feature probably looks something like Figure 4(b), where switch *S1* is in the LEC system serving telephone *LT*, and switch *S2* is in the IEC system. As explained in Section 3.3, when the path request reaches the IEC system at *S2*, the IEC system discovers that it is a request for credit (probably because of the dialed string, which is part of the *request-path* signal), and needs to collect a credit account number. It completes the voice path to *LT*, prompts for and collects the number through in-band signaling, and then (if the account number is good) extends the path to *RT*. Meanwhile the LEC system serving *LT* has nothing to do with the credit-card feature, and need not even detect that it is being used.

Also consider a conference among three telephones accessed directly by systems *B*, *C*, and *D* respectively. *The conference feature could be provided by any one of the systems A, B, C, or D.* The bridge is located in the system providing the conference feature, and the other systems see nothing more than plain voice paths.

In a typical conference, one telephone plays the role of the controller, and is the only device with the power to add parties, drop parties, or transfer (in Figure 3 the left telephone is the controller). Thus the controlling telephone must have a way of transmitting conference-control signals to the system providing the conference.

If the controlling telephone and the conference-providing system have a direct-access relationship, then the solution to this problem centers on their line protocol, which must include signals for conference control. For example, if it is an analog line, then an existing signal such as a *flash* will acquire a special meaning in the conference context (this is how the *three-way calling* feature works [5]).

If the controlling telephone and the conference-providing system have an indirect-access relationship, on the other hand, then in-band signaling is the only possibility, because intersystem protocols do not support conference control. In this case a voice-processing device will be attached, by the conferencing system, to the path from the controlling telephone. Because the device will be attached in *monitor mode,* it will monitor the voice input from the controlling telephone for touch-tones or keywords, without interrupting the voice path in any way.

5.   The Call-Processing Interface

This section concerns the interface that the customer-oriented telephony machine uses to control call processing in the voice-transmission machine, as shown in Figure 1.

When we are concerned with a single telephone system, we are dealing with a vertical slice of Figure 1, as shown in Figure 6. The customer-oriented telephony machine lies within the system and is isolated from other telephone systems. The lateral interfaces of the voice-transmission machine are the lines and external trunks by which it is connected to the rest of the world-wide network, including both their voice channels and associated signaling channels (if any).

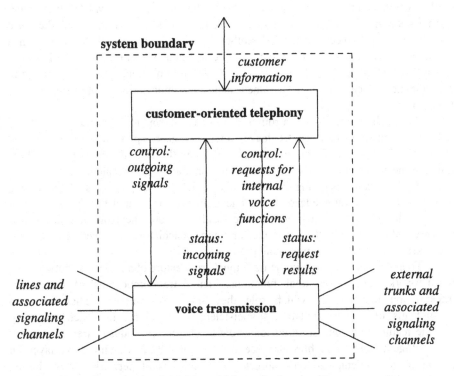

Figure 6. The call-processing interface within a telephone system.

The voice-transmission machine reports to the customer-oriented machine signals received from lines and external trunks. Some of the commands from the customer-oriented machine to the voice-transmission machine instruct the latter to send out control signals on the lines and external trunks. The other instructions from the customer-oriented machine are requests for voice functions internal to the system.

As you might expect, the perspective needed for the customer-oriented layer is quite different from the natural perspective of the voice-transmission layer. Looking inward from the lines and external trunks, the primary concern of the customer-oriented machine is to bring about *connections* among sets of lines and external trunks. A connection is a relationship whose presence enables mutual voice communication. The primary concern of the voice-transmission machine, as we have seen, is voice paths. Voice paths are different from connections, for the following two reasons.

(1) Connections are always local to one telephone system. Making connections is the main thing that a telephone system does for its customers, and the system must have some control over all the ingredients used to make them. Voice paths, in contrast, are global. They can pass to or through telephony devices and external switches, neither of which is under the system's control.

(2) Connections describe telephony at a higher level of abstraction than voice paths do. For example, at the level of the customer-oriented machine, a three-way conference is simply a ternary connection relation. At the level of the voice-transmission machine, it is implemented using a conference bridge and the local sections of three voice paths. For another example, a voice path leading from a telephony device to a voice-processing device within a system might not be involved in any connection made by that system. It might be implementing a user interface to one telephone (see Section 6.2) rather than a communication relationship among telephones.

Although Figure 6 gives a rough idea of the information that must pass across the call-processing interface, it leaves many choices open: what the level of abstraction is, how the functions are grouped, how the information is filtered, how much of the architecture of the voice-transmission machine is revealed, etc.

A clean call-processing interface is desirable for two reasons. One reason, which is widely acknowledged in the telecommunications industry, is the need to upgrade the voice-transmission machine easily without incurring the delay and expense of changing the customer-oriented machine. As everyone knows, transmission technology is improving rapidly.

To elaborate this point, in the traditional architecture of a telephone network, all the network nodes have dual functions. They are components of the voice-transmission machine, in which role they serve as switches. They are also components of the distributed customer-telephony machine, in which role they provide feature logic and store customer data. The current industry trend is toward the Intelligent Network architecture (see also Section 6.3), in which the two layers in Figure 6 are implemented on completely disjoint sets of network nodes. In the Intelligent Network architecture, the switches are concerned only with voice transmission and related functions. The customer-oriented telephony machine is implemented completely within other network nodes specializing in feature logic and customer data.

The other reason for a clean call-processing interface, which is less widely understood, is pure separation of concerns. Each of the layers is subject to many pressures toward change, growth, and increased complexity. When the two layers intertwine, an increase in the complexity of one tends to increase the complexity of the other, enabling a feedback loop with very unfortunate consequences.

6. Observations

6.1. On Distribution

Most specifications of telephony features assume that the voice-transmission machine is centralized without justifying the assumption or even mentioning it (e.g., [20]). Some specifications deal honestly with the fact that it is not, at great cost to their writers and readers (e.g., [17]).

The stark reality is that the features of real telephone systems are extremely complex. At present we cannot specify them completely even with the centralization assumption, let alone without it.

It makes sense to use the centralization assumption for feature specification, even when the voice-transmission machine is distributed and the assumption is false. Techniques such as those suggested by Jacob [13] might make it possible to specify a distributed system as if it were centralized, while maintaining a suitable formal relationship between the specification and the thing specified. Even without the formal relationship, a specification based on this assumption is much better than no specification at all. Telephone engineers are very well accustomed to working with imperfect specifications, and implementing them to "reasonable and customary" standards.

Use of the centralization assumption has another valuable consequence: telephone systems in which the voice-transmission machine is centralized and in which it is distributed can both be specified using the same techniques.

6.2. On User Interfaces

The "user interface" of telephony consists of what a person using a telephony device hears, sees (on the device's displays), and can do to choose or affect services. A telephone system implements a particular user interface by emitting and accepting certain signals associated with lines and external trunks. This section consists of three simple observations about these interfaces.

(1) Every telephone system has to provide more than one user interface. Systems with direct-access interfaces must also provide indirect-access interfaces. Systems with only indirect-access interfaces always have external trunks with different protocols. Wireless access has some characteristics of indirect access and some of direct access; the cellular systems that provide wireless access must also provide indirect access. Despite this variety of interfaces, I have never yet seen a paper on telephony specification even mention multiple interfaces, let alone provide them.

(2) In specifying user interfaces, it is extremely important to unify in-band and out-of-band signaling. Both are heavily used, and they are used for overlapping functions. Despite this obvious fact, I know of only one paper that even mentions in-band signaling (it includes a proposal for unifying the two kinds of signaling in interface specifications [27]).

(3)   The literature on protocol specification may seem relevant, but the protocol approach is not necessarily the best approach to telephony.

In the protocol literature, there is a "constraint-oriented style" of specification in which the specification is decomposed into local (endpoint) and end-to-end constraints [3,21,23]. This style has been used for telephony specification in LOTOS [8]. In the LOTOS specification, the local constraint is a description of the user interface for a line. The end-to-end constraint is a description of the relationship between two interfaces brought about by the telephone system for the purpose of connecting their lines.

An important fact about the constraint-oriented style is that the local constraints are pure projections of the end-to-end constraints [3]. In other words, the local constraints can be derived from the end-to-end constraints. This makes sense for protocol specification, where end-to-end constraints are arguably the only important properties, but it does not make sense for telephony because of the following telephone capability.

Some telephone systems provide user interfaces for other purposes than for establishing connections. Interactive voice interfaces enable customers to update databases and retrieve information. Interactive voice interfaces are also used to collect information such as authorization codes, directory numbers, and credit account numbers as part of screening, routing, and billing features, respectively. Note that these latter interfaces are active before any attempt to connect is made.

Thus the behavior of a user interface is often independent of any connection attempt. By allowing it to stand on its own, we increase separation of concerns and decrease redundancy. While it is true that the implementation of some interfaces involves voice paths that extend into a telephone system to reach voice-processing devices deep within its network, as explained in Section 5, that is a detail with no place in a specification.

When interfaces are treated realistically, the specifier often faces the complexity of a many-to-many mapping between interface and connection events. Some specification techniques have been developed to manage this complexity [28].

6.3. Calls Considered Harmful

A *call* is an attempt by one telephone (the *caller*) to establish one connection to one other telephone (the *callee*). The state of the call encodes or implies information about the states of all three entities. A *call model* is a conceptual model for telephony that describes all telephony in terms of calls.

The most advanced call model in use today is the Intelligent Network Conceptual Model (INCM). It was developed under the auspices of the International Telecommunication Union (ITU) and the European Telecommunication Standard Institute (ETSI), and is being promulgated as a standard by those organizations [7,9,12,15]. The primary purpose of the INCM is to support evolution toward the Intelligent Network architecture, as introduced in Section 5.

The INCM defines points in a call where control of the call can pass from a switch to a feature node and back again. A few control point names and definitions [15] show the nature of the INCM:

Address Collected: This point identifies that the destination address has been received from the user.

Busy: This point identifies that the call is designated to a user who is currently busy.

Active State: This point identifies that the call is active and the connection between the calling and called parties is established.

Clearly these control points are combined states of a caller interface, a connection or potential connection, and a callee interface or potential callee.

Some features fit into the call model very well. Here are some examples of features and capabilities that the INCM is designed to handle.

Translation. Translation from one directory number to another is used for forwarding, 800/freephone calling, and other services. It can occur at any point before an attempt to connect is made.

Screening. Screening features are used to authorize and deny calls. Like translation features, they can be invoked at any point before a connection attempt is made, and have no effect on later stages of the call.

Queueing. Some callers compete for the attention of a pool of service agents. Calls are queued, and connected to available agents in FIFO order. Queueing can be represented as a nondeterministic preconnection delay in the history of a call.

Many other features and telephone capabilities, however, subvert the one-to-one-to-one correspondence that is the essence of the call model. Up to a point, the call model can be extended (patched) to accommodate these features. Beyond that point, the call model is likely to break down in hopeless complexity. The following examples present some of the features and capabilities that most deeply undermine the assumptions of the call model.

Conferencing. Conferences connect sets of interfaces, from three to a hundred. The call model, of course, is concerned with only two interfaces. For a large conference, an appointment must be made ahead of time. Thus, at its starting time, a large conference can be thought of as a connection with no interfaces. Furthermore, the conference can even initiate creation of the voice paths to all the participating telephones—very different from a call, which is always initiated by a caller.

Serial and time-multiplexed connections. A credit-card customer enters account information, then proceeds to make a series of calls on the same account before hanging up (he presses "#" on the dialpad to end a call without hanging up). In this situation, one interface makes a series of independent connections. Alternatively, customers with call waiting or multibutton telephones can time-multiplex several independent connections from within the same interface.

Delayed communication. A voice message is like a call in the sense that there is a caller and a callee, and the callee hears what the caller says. But the caller and callee are never connected to each other, and are not even accessing the system at the same time! Furthermore, the telephone system offering the messaging service might call or be called by the recipient of the message. Automatic-callback features also introduce multiple communication phases, separated by time.

Interface-only features. As discussed in Section 6.2(3), many telephone features provide user interfaces without or before any attempt to make a connection.

Indirect access. If we take seriously the idea that a call connects telephones (rather than lines or trunks), then a long-distance call is a multi-system phenomenon, and an IEC system (providing only indirect access) can never complete a call on its own.

These counterexamples suggest to me that the call model is as much of a hindrance as a help in understanding telephony. It is easy enough to see how its dominance arose. Partly it is the overwhelming historical influence of POTS. Partly it is the presence of an important point-to-point concept in telephony—but the concept is the voice path, which belongs at the physical level and not at the logical level. And partly it is the immediate practical problem that we have nothing better to put in its place.

6.4. How Formal Methods Can Help: Feature Interaction

Telephone engineers do not need formal methods to help them implement POTS: there have been successful implementations of POTS for about 75 years. These final two sections describe aspects of telephony in which help is needed, and for which the help needed seems to be the kind that formal methods can provide.

Most generally, the feature-interaction problem [5,11,25] is the problem of making a telephone system behave the way we want it to despite continual, incremental expansion of services. This has proved to be a very difficult problem; despite much attention from researchers it has scarcely been alleviated.

There is a range of formal approaches to the feature-interaction problem. At one end of the range is the detection approach, in which features are specified independently in a compositional language. The composed features are analyzed algorithmically to detect inconsistencies and failure to satisfy desired properties (e.g., [1,2,4,16,20]).

At the other end of the range is the structural approach, emphasizing modular specification and separation of concerns. The idea is to organize the specification so that desirable properties are guaranteed by its structure, and so that it is easy to add features without destroying its structure or exploding its complexity. The notion that features can or should be specified independently receives less emphasis than in the detection approach (e.g.,[6,14,26,29]).

These two approaches have complementary advantages and disadvantages. The following two comparisons capture the most important points.

(1)   Detection research is more straightforward to carry out. A researcher can simply apply an existing language and analysis tool to the problem and see what happens. In-depth knowledge of telephony is not usually needed and seldom influences the results.

Structural research, on the other hand, is groping in the darkness. Researchers cannot assume that existing languages and tools are adequate. The more knowledge of telephony available, the better. As a result of all these factors, structural research often leans heavily on the structure of the implementation, thus compromising the call-processing interface.

(2)   In a pungent critique of detection research [22], Velthuijsen observes that no one has yet succeeded in using algorithmic analysis to detect a major feature interaction that was not previously known. The reason is that features are almost

never orthogonal—in almost all cases, adding a feature creates exceptions and requires changes to the previously specified features. The goal property checked by the algorithm must incorporate all of the exceptional cases. By the time a person has written the property correctly, he already understands all of the exceptions and potential interactions, both desirable and undesirable.

Even if the detection approach succeeds to perfection, how can the specification errors be corrected so as to produce a well-structured, readable specification of the whole system? My experience suggests that the corrections will form a cascade of ugly and unmanageable exceptions.

The structural approach does not suffer from these disadvantages. Rather, its whole focus is to avoid them by eliminating exceptions and providing a readable overall specification.

These two comparisons show that there is no clear winner. Nor does there need to be, since the two approaches can be combined (in fact, many of the examples cited mix elements of both). Nevertheless, I believe that the structural approach is more fundamental and more necessary than the detection approach. The structural approach is the one that seeks to discover and exploit knowledge of the application domain.

There is one other possible approach that deserves some attention. It would be very helpful to have a robust collection of simple, abstract properties of well-behaved telephone systems ("principles of telephone etiquette"). For example, consider the principle, "A subscriber is never billed for a call unless he knows he is paying for it." This principle would be violated by a system that sets up collect calls without informing the callee. This principle might also be violated through a call to an 800/freephone number, if the 800 number translates to a normal 900 number.[7]

Such principles are difficult to find, as Velthuijsen's remarks make clear. Success would require working at a much more abstract level than the properties used for detection, formalizing vague concepts such as "what a subscriber knows he is paying for." Success would also require giving the principles prescriptive, as well as descriptive, force, since some current features are sure to violate them.

Such principles would support the other approaches in obvious ways. For one example, they could be checked by detection mechanisms. For another example, they could be used to constrain or derive the detailed behavior of features within the boundaries of a structural approach. But the hierarchy of abstract telephony concepts needed to construct the principles would provide a great deal of insight and organization in their own right.

6.5. How Formal Methods Can Help: Separation of Concerns

The clean call-processing interface described in Section 5 is more of a goal than a reality. There is widespread eagerness to find a specific interface that will separate the concerns of these two layers effectively while preserving maximum flexibility on both sides.

---

[7] In addition to the transmission cost of the call, a caller to a 900 number also incurs a per-minute charge payable to the callee.

Acknowledgments

I am indebted to Michael Jackson for many years of fruitful discussions about telephony. Anthony Finkelstein, Daniel Jackson, Gerald Karam, Nils Klarlund, and Jim Woodcock also provided helpful comments.

References

[1] Johan Blom, Roland Bol, and Lars Kempe. Automatic detection of feature interactions in temporal logic. In K. E. Cheng and T. Ohta, eds., *Feature Interactions in Telecommunications Systems III*, pages 1-19. IOS Press, 1995.

[2] Johan Blom, Bengt Jonsson, and Lars Kempe. Using temporal logic for modular specification of telephone services. In L. G. Bouma and H. Velthuijsen, eds., *Feature Interactions in Telecommunications Systems*, pages 197-216. IOS Press, 1994.

[3] Gregor V. Bochmann. A general transition model for protocols and communication services. *IEEE Transactions on Communications* XXVIII(4):643-650, April 1980.

[4] Kenneth H. Braithwaite and Joanne M. Atlee. Towards automated detection of feature interactions. In L. G. Bouma and H. Velthuijsen, eds., *Feature Interactions in Telecommunications Systems*, pages 36-57. IOS Press, 1994.

[5] E. Jane Cameron, Nancy D. Griffeth, Yow-Jian Lin, Margaret E. Nilson, William K. Schnure, and Hugo Velthuijsen. A feature interaction benchmark for IN and beyond. In L. G. Bouma and H. Velthuijsen, eds., *Feature Interactions in Telecommunications Systems*, pages 1-23. IOS Press, 1994.

[6] D. Cattrall, G. Howard, D. Jordan, and S. Buj. An interaction-avoiding call processing model. In K. E. Cheng and T. Ohta, eds., *Feature Interactions in Telecommunications Systems III*, pages 85-96. IOS Press, 1995.

[7] José M. Duran and John Visser. International standards for intelligent networks. *IEEE Communications* XXX(2):34-42, February 1992.

[8] Mohammed Faci, Luigi Logrippo, and Bernard Stepien. Formal specification of telephone systems in LOTOS: The constraint-oriented style approach. *Computer Networks and ISDN Systems* XXI:53-67, 1991.

[9] James J. Garrahan, Peter A. Russo, Kenichi Kitami, and Roberto Kung. Intelligent Network overview. *IEEE Communications* XXXI(3):30-36, March 1993.

[10] J. A. Goguen and Luqi. Formal methods and social context in software development. In *Proceedings of the Sixth International Conference on Theory and Practice of Software Development (TAPSOFT '95)*, pages 62-81. Springer Verlag LNCS 915, 1995.

[11] Nancy D. Griffeth and Yow-Jian Lin. Extending telecommunications systems: The feature-interaction problem. *IEEE Computer* XXVI(8):14-18, August 1993.

[12] ITU-T/ETSI Recommendations Q1201-Q1205, Q1211, Q1213, Q1214. 1993.

[13] Jeremy L. Jacob. Refinement of shared systems. In John McDermid, editor, *The Theory and Practice of Refinement: Approaches to the Formal*

Development of Large-Scale Software Systems, pages 27-36. Butterworths, 1989.

[14] Yoshiaki Kakuda, Akihiro Inoue, Hiroyuki Asada, Tohru Kikuno, and Tadashi Ohta. A dynamic resolution method for feature interactions and its evaluation. In K. E. Cheng and T. Ohta, eds., *Feature Interactions in Telecommunications Systems III*, pages 97-114. IOS Press, 1995.

[15] Jalel Kamoun. Formal specification and feature interaction detection in the Intelligent Network. Department of Computer Science, University of Ottawa, Ottawa, Ontario, 1996.

[16] Yasuro Kawarasaki and Tadashi Ohta. A new proposal for feature interaction detection and elimination. In K. E. Cheng and T. Ohta, eds., *Feature Interactions in Telecommunications Systems III*, pages 127-139. IOS Press, 1995.

[17] Andrew Kay and Joy N. Reed. A rely and guarantee method for Timed CSP: A specification and design of a telephone exchange. *IEEE Transactions on Software Engineering* XIX(6):625-639, June 1993.

[18] Evan H. Magill, Simon Tsang, and Bryce Kelly. The feature interaction problem in networked multimedia services: Past, present and future. EPSRC No. GR/K 72995, October 1996, http://www.comms.eee.strath.ac.uk/~fi/fimna.html.

[19] N. Mitra and S. D. Usiskin. Relationship of the Signaling System No. 7 protocol architecture to the OSI Reference Model. *IEEE Network* V(1):26-37, January 1991.

[20] Tadashi Ohta and Yoshio Harada. Classification, detection, and resolution of service interactions in telecommunication services. In L. G. Bouma and H. Velthuijsen, eds., *Feature Interactions in Telecommunications Systems*, pages 60-72. IOS Press, 1994.

[21] K. J. Turner and M. van Sinderen. LOTOS specification style for OSI. in Jeroen van de Lagemaat and Tommaso Bolognesi, editors, *The LOTOSPHERE Project*, pages 137-159. Kluwer Academic Publishers, 1995.

[22] Hugo Velthuijsen. Issues of non-monotonicity in feature-interaction detection. In K. E. Cheng and T. Ohta, eds., *Feature Interactions in Telecommunications Systems III*, pages 31-42. IOS Press, 1995.

[23] Chris A. Vissers, Giuseppe Scollo, Marten van Sinderen, and Ed Brinksma. Specification styles in distributed systems design and verification. *Theoretical Computer Science* LXXXIX(1):179-206, 1991.

[24] Pamela Zave. Application of formal methods is research, not development. *IEEE Computer* XXIX(4):26-27, April 1996.

[25] Pamela Zave. Feature interactions and formal specifications in telecommunications. *IEEE Computer* XXVI(8):20-30, August 1993.

[26] Pamela Zave. Secrets of call forwarding: A specification case study. In *Formal Description Techniques VIII (Proceedings of the Eighth International IFIP Conference on Formal Description Techniques for Distributed Systems and Communications Protocols)*, pages 153-168. Chapman & Hall, 1996.

[27] Pamela Zave and Michael Jackson. Requirements for telecommunications services: An attack on complexity. In *Proceedings of the Third IEEE*

International Symposium on Requirements Engineering, pages 106-117. IEEE Computer Society Press, 1997.

[28] Pamela Zave and Michael Jackson. Where do operations come from? A multiparadigm specification technique. *IEEE Transactions on Software Engineering* XXII(7):508-528, July 1996.

[29] Israel Zibman, Carl Woolf, Peter O'Reilly, Larry Strickland, David Willis, and John Visser. Minimizing feature interactions: An architecture and processing model approach. In K. E. Cheng and T. Ohta, eds., *Feature Interactions in Telecommunications Systems III*, pages 65-83. IOS Press, 1995.

Convergence of Telecommunications and Computing on Networking Models for Integrated Services and Applications

Maurizio Decina*

Politecnico di Milano/CEFRIEL
Milano (Italy)

Abstract

This paper gives a snapshot on the convergence of telecom and computer industries on a common vision on networking models against the stringent requirements of multimedia applications on future integrated services information infrastructure.

At first, we summarize requirements and elements of such infrastructure, and introduce a networking model that is framed into four layers: interconnection (bearer), internetworking, interoperability (middleware), and application. Then, we offer a brief review on the classical telecom networking model, the current status of ATM multimedia networking features, and the ongoing activities for fostering evolution of telecom networks to enhance control, management and service delivery. An overview on the current Internet network architecture follows, and the recent achievements and the evolving networking protocols to cope with real-time multimedia applications are presented.

The described integration schemes of IP and ATM resource reservation protocols, to provide efficient resource utilization and QoS guarantee in broadband networks, fall into the first two lower architectural layers, which are the foundation of network computing and communication services. In particular, we present a summary description of the comparative features of both IP and ATM protocol approaches, and we report on current proposals for handling IP on ATM networks.

Finally, we briefly discuss trends in middleware services, according to the emerging technology of transportable computation and intelligent agents. Interoperability of heterogeneous networked systems is based on an agreed distributed program encoding and computation environment, thus pushing the key point of the layered networking architecture up into the middleware layer.

---

* Prof. Maurizio Decina, Centro Cefriel - Politecnico di Milano, Via Emanueli, 15 - 20126 Milano- Italy, Tel: +39 2 66100643, FAx: +39 2 66100448, http://www.cefriel.it/ decina, decina@mailer.cefriel.it, m.decina@ieee.org

The NetAcademy –
A New Concept for Online Publishing and
Knowledge Management

Siegfried Handschuh, Ulrike Lechner, David-Michael Lincke, Beat Schmid,
Petra Schubert, Dorian Selz, Katarina Stanoevska-Slabeva

Institute for Media and Communications Management, University of St. Gallen
Müller-Friedberg-Strasse 8, CH–9000 St. Gallen, Switzerland
EMail: `Firstname.Lastname@mcm.unisg.ch`

Abstract. Traditional media have concepts to ensure quality of information they carry, while new media make information ubiquitious. The NetAcademy project constitutes a new medium for knowledge accumulation and dissemination for scientific purposes. It provides by its underlying carrier, the Internet, access to information and by its management concepts quality of information.
We explore the NetAcademy with its open, distributed architecture, the NetAcademyNet and discuss how such a medium as the NetAcademy will influence the process of publishing and scientific work.
Keywords: Online publishing, Knowledge Management, Multi-agent system.

1   Introduction

Traditional media and channels for dissemination of knowledge, like, e.g., libraries, journals and conferences have concepts and techniques to ensure quality of information they carry. In addition, they provide classification, retrieval and query mechanisms, which guarantee relevance and restrict hereby the quantity of search results.

New media and channels such as the Internet have advantages compared to traditional channels. Information can be published without delay and can be retrieved (nearly) instantly throughout the world at any time, independent of the physical location of the medium at which it is stored. In addition, related information can easily be linked and made available as a whole. E.g., references from a scientific publication available on the net, can be linked to the publication. Thus, information relevant for the publication is available on fingertips.

New media open up opportunities for scientific publishing. Scientific results can be made available instantly and worldwide and, moreover, publications can be narrowed down to the core statements and linked with related publications available on the medium [Giu97].

The easiness with which information can be published on the Internet has given rise to publishing on this medium. But despite the above described potentials of the new medium the process of scientific publishing has not significantly changed. The new medium is used like the conventional ones - linear text is made available without qualified links to related and referenced work. In some aspects the use of the new medium has even worsened the quality of scientific publication as it provides an easy way of publishing information which is neither reviewed nor accepted by a scientific community. In addition, there are no qualified and adequate classification and search mechanisms for scientific results. By searching for contributions related to a certain topic, the range of answers goes from abstracts presenting somebodies thoughts to papers submitted in a journal or conference proceedings and to copies of already reviewed and published papers. In general, information search and retrieval via the mechanisms provided on the Net is not satisfactory with respect to the quantity and, in particular, the quality of information; Filtering relevant, high quality information from the result of a query of a search engine is time-consuming and cumbersome.

The above described current state of publishing on the Internet is not acceptable especially for scientific publication. Quality, mutual reference and linkage of information as well as quality and relevance of search are necessary requirements for scientific publishing. Therefore, scientific communities need media and channels to accumulate and disseminate knowledge with the characteristics and advantages of traditional as well as new channels: quality and speed. The NetAcademy is designed to provide both. It is a platform for a scientific community providing through the technology of its medium, the Internet the speed for publishing and the accessibility of information (nearly) independent of space and time. It provides also the management facilities to ensure the quality of the information stored as well as the quality of the information retrieval mechanisms.

The medium and the applied process of information accumulation and dissemination influence themselves mutually. Thus, the NetAcademy as a medium has the potential to change and renew scientific publishing and knowledge management within scientific communities. Let us illustrate the influence of the medium NetAcademy. The Internet technology provides means to link information and the NetAcademy concept provides an organizational framework and an information space of qualified contributions together with classification and retrieval mechanisms. Small units of information can be published, comprising only essential information and pointers to existing information like, e.g., the basic definitions and the context. The NetAcademy as a platform fosters communication as well as inter-disciplinary research and facilitates by its underlying technology distributed cooperative authoring. It applies and adopts an electronic counterpart to well known and trustworthy conventional review processes for scientific publications. Thus, management processes of knowledge creation, approval and dissemination will change as well. They will speed up since all the communication can be done on the Internet and, moreover, they can be partly automated. The technology of the NetAcademy ensures, e.g., the origin and the accuracy of other data on the information stored in a NetAcademy, and thus, the information in a NetAcademy is trustworthy, which is an important quality criterion.

A NetAcademy is a platform for a single scientific community. The NetAcademy concept provides ways to relate different NetAcademies. Information can be retrieved across different NetAcademies based on mechanisms to mediate between the different terminology employed by the single NetAcademies. Thus, the NetAcademies form—in analogy to their underlying medium—a NetAcademyNet. This fosters inter-disciplinary research as well as knowledge sharing.

We present in this paper the basic concept of the NetAcademy and the current state of its implementation. The focus of the paper lies on the concepts for organization and retrieval of knowledge.

This paper is organized as follows. Sect. 2 presents the concept of representing and organizing knowledge and introduces briefly the terminology of knowledge media. Sect. 3 describes the concept of a Net Academy Net. Sect. 4 contains a brief description of the already established NetAcademies and their contents. Sect. 5 gives an overview over the implementation of the NetAcademy platform, the user interface (Sect. 5.1) as well as the architecture and technology (Sect. 5.2). Related work is discussed in Sect. 6.

2 Knowledge Representation and Management

In this section, we explain the concept of the NetAcademy [Sch97], define the basic terminology and describe the representation, organization and management of knowledge in a NetAcademy. In order to achieve this, first we explain the generic concept and terminology of knowledge media - the underlying theoretical concept of the NetAcademy. Based on this description, in a second step, we instantiate a general template for a NetAcademy and finish this section with a more precise description of an applied concept of the Net Academy.

2.1 Knowledge Media: Knowledge, Agents, Channels and Media

The NetAcademy concept draws from a generic concept on knowledge media that we briefly introduce in this section. For a more detailed explanation we refer to [Sch98].

A *knowledge medium* is an information space, which supports knowledge exchange within a community of agents. A *knowledge medium* comprises beside *agents*, an organizational structure consisting of a collection of *locations for agents*, *roles*, *protocols* and *processes*, a collection of *channels*, a *logical framework*, and a class of *worlds*, to which the formal framework refers.

Agents are active, autonomous and communicating entities, with locations that differ in space and time. They are the basic source of information and knowledge in the medium. They manipulate knowledge and trigger changes in the knowledge medium.

Knowledge is an agent's subjective view of a world. It is the internal representation of the world, which an agent is part of. Agents gain knowledge either by observation, or by communication with other agents. Knowledge comprises data as well as behavior, a language to code it and a calculus for inference of conclusion. In order to express and communicate knowledge, agents externalize knowledge in terms of coded information. To facilitate communication, externalized *knowledge* has to be represented by some code on a carrier, the *channel*. Channels are entities capable of carrying and transporting knowledge. They connect the locations of agents as a means for agents to communicate, i.e., to bridge differences in space and time. Knowledge is represented in channels by a *logical framework*, with some (formal) language for representation of information and a calculus for inference. The logical framework is agreed upon by agents and is the base for mutual understanding, i.e., for the appropriate communication of semantics of knowledge.

Knowledge generation, management and dissemination is performed within knowledge media in a defined *organizational structure*. Agents form or are grouped in communities striving towards a common goal or representing a common view, on a certain domain of discourse. Each community constitutes *roles*, which are taken over by agents and which determine the rights to access information as well as the behavior of agents. *Processes* are formalized *protocols* of the activities, which take place within the medium and can be defined over roles of agents. Core processes supported by a knowledge medium are the ones for submitting content to the medium, quality assurance of contributions, retrieval of content from the medium as well as management processes for knowledge as deletion, copy etc.

Knowledge Media can be structured hierarchically. A knowledge medium can be viewed as an agent that is part of another knowledge medium.

The basic components of a knowledge medium, the agents, channels, roles, protocols and the logical framework, have different features depending on the characteristics and requirements of the target communities and the possibilities of technologies used for their realization. In the next section, we describe the instantiation of this concept by the NetAcademy, a knowledge medium for scientific communities.

2.2 The NetAcademy as a Knowledge Medium

A NetAcademy is an instance of a knowledge medium for the scientific community. Taking into consideration the specific requirements of scientific communities and based on the concept knowledge medium we explore a general template for a knowledge medium and instantiate it with the NetAcademy. The core components of the template and their interrelationships are depicted in Fig. 1.

Agents of a scientific community are researchers, which have a common domain of discourse and language, as well as all other people interested in the topic of a scientific community. Thus, agents in a NetAcademy are either representations of natural persons as, e.g., scientists participating in the NetAcademy or systems capable of performing functions related to knowledge generation and dissemination. Agents of a NetAcademy can take different roles, which are aligned on the currently prevailing scientific review and publishing processes. They can be reviewers, editors, chief editors, authors or just consumers of the offered content.

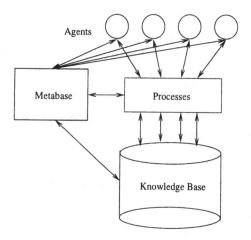

Fig. 1. NetAcademy-template

At its core, a NetAcademy contains a *knowledge base*. The knowledge base stores externalized knowledge in channels typical for a scientific community as, e.g., facts and procedural knowledge, discussions, research papers and other information relevant to the domain. It comprises research results concerning theoretical issues as well as practical applications.

Besides channels containing core results of scientific work, another important part of externalized knowledge is its classification and meta-description. This task is performed in a NetAcademy by the metabase. The *metabase* includes the vocabulary of a NetAcademy. This *vocabulary* represents the terminology employed and represented in the knowledge base and establishes a meta-level by which the knowledge can be accessed and by which NetAcademies can communicate.

The *process* layer formalizes the protocol, i.e., the rules for the agents accessing the knowledge base. Processes implement the protocols, according to which the knowledge base may be accessed and manipulated, and they implement (partly) automated management tasks. In the case of the NetAcademy such processes are: submitting contributions to the knowledge base, review of submitted contributions and management processes of the content.

Based on the above described general template for a scientific knowledge medium a specific knowledge medium - The NetAcademy - was designed and implemented by using specific approaches for organization and retrieval of knowledge.

The information stored in the NetAcademy comprises typically a lot of publications, part of which have been published elsewhere, as well as publications exclusively published in this NetAcademy. It may contain a number of references to publications, as well as documents describing open problems or hypotheses and information on conferences, their deadlines and their programs and pointers to other information bases. There might be ongoing discussions, whose (agreed) results are archived in the knowledge base. Publications are papers—like in conferences or journals—as well as short notes and sketches. A knowledge base typically contains meta information, like ratings of the pieces of information, e.g., whether they are basic knowledge or cover advanced topics, whether they have theoretical or practical flavor.

The content of a NetAcademy is classified in two major parts: a section of reviewed contributions, with a NetAcademy *stamp of quality* and a part containing contributions submitted for reviewing or being in the review process. The NetAcademy stamp of quality is an information set containing information about the author, date of submission as well as about the review process, which the publication has gone through.

The meta-description is given as a predefined hierarchical structure [Sta97]. It provides the interface by which semantic (inter-NetAcademy) query mechanisms access the knowledge base (See below and Sect. 3.2).

The NetAcademy provides two query mechanisms to access the information in the data base. A conventional syntax based one and a semantic one. The syntax based one employs the technology of typical Internet search engines. Let us explain the semantic query mechanism. It uses the vocabulary to access the database. The vocabulary is organized in a hierarchy, where terms upper in the hierarchy are more general than terms lower in the hierarchy. Let T be the term used in a query, and let $S_1, \ldots, S_n$ be specializations of T, such that $S_1, \ldots, S_n$ are lower in the vocabulary's hierarchy. A query for information concerning T yields all information concerning T plus all information concerning $S_1, \ldots, S_n$. Thus, the semantic query mechanism allows to abstract from the terminology.

Reviewing is a process of rating and ensuring the quality of information in academia. Currently, it mirrors the prevailing review process of scientific journals. Part of this process can be automated as, e.g., sending a paper submitted to the reviewers, collecting the reviews or reminding the reviewers. To ensure the quality of the query facilities, the scientific community may decide to lay out the process of publishing on the NetAcademy such that the author is obliged to provide the vocabulary for the publication, and–if necessary–to extend the vocabulary.

A lively NetAcademy earns the denotation "knowledge base", since it constitutes a representation of the universe of discourse of a scientific community and an image of the knowledge of its members. Like and with the knowledge of the members of that community, it will grow, change and evolve and it provides them with a means of communication that traditional carrier systems and organizations of information do not provide.

3   The NetAcademyNet

A single NetAcademy is a platform for organizing knowledge and management for a (scientific) community. According to the scientific community, whose knowledge it represents, it is not isolated, but embedded in a net of communicating

NetAcademies, whose knowledge is related. Thus, NetAcademies are considered to be agents in the knowledge medium NetAcademyNet.

3.1 Relations between NetAcademies

There are three relations between NetAcademies, determined by the way, how the knowledge they contain is related.

- *Information sharing.* NetAcademies with similar theories represent non-disjoint worlds. Some of the knowledge they contain refers to the same domain. An example for such a sharing of knowledge are, e.g., knowledge bases on object-oriented programming and Pascal programming, sharing information about basic algorithms and programming techniques.
- *Inter-disciplinary.* NetAcademies provide different views on the same world, i.e., they represent knowledge about the same universe of discourse, but employ different theories, to model and represent it. An example are two NetAcademies dealing with media. One employs computer science to explain media by their implementation using computer science terminology, one using business science to explain the motivation, the process of making money on the Net, and one of media science, dealing with the look and feel, with the characteristics of media.
- *Generalization and specialization.* There might be a hierarchy of NetAcademies, on one topic, i.e., a NetAcademy on Computer Science, containing basic terminology, techniques and methods and several NetAcademies on special issues, like, e.g., concurrency, object orientation dealing with smaller worlds, employing specialized vocabulary.

These three relations are based on the world, i.e., the domain of discourse a NetAcademy deals with. To facilitate communication, i.e., exchange of information between NetAcademies, one has to abstract from the theory and its terminology. The vocabulary is here the interface. The vocabulary of a NetAcademy contains the terminology of a NetAcademy organized in a hierarchy allowing semantic access to the knowledge base. The abstraction of the vocabulary and the mediation between different vocabularies is explained in Sect. 3.2.

This Net of NetAcademies has a global root, called the *NetAcademy on Net-Academy*, which acts as a central entry point to the distributed NetAcademy knowledge medium and offers a global directory of all NetAcademies. As a meta-layer its contents present the theoretical foundation of the NetAcademy and the meta-information on knowledge.

3.2 Communication between NetAcademies

The vocabulary as part of the meta base is—with its hierarchical structure—the interface to the contents of the knowledge base. Accordingly, we abstract from the terminology and thus from the particular theory a NetAcademy and its scientific community employs.

We employ the Q-language as this abstract language, common to all the NetAcademies, and the Q-calculus for reasoning [SGW⁺96,Sta97].

The language of the Q-calculus comprises sorts with attributes, whose values belong to scales, transitions and transition classes. It provides inheritance and constraints to define sub-sorts.

The abstraction from a vocabulary of a NetAcademy to a common vocabulary in the Q-language assigns an abstract sort with values of attributes to each term of the concrete language.

Let us describe, how the search across NetAcademies with their vocabularies works. Assume we have a query in NetAcademy N. N generalizes a collection of NetAcademies $A_1 \ldots A_m$ and provides a view of a world, for which the NetAcademies $B_1 \ldots B_n$ also provide views. We would like to to search for information in N (which includes by generalization $A_1 \ldots A_m$) as well as $B_1 \ldots B_n$.

A query S phrased in the vocabulary in one NetAcademy N is translated to the abstract level of the Q-calculus, mapped to a number of (concrete) queries $S_N, S_{A_1}, \ldots, S_{A_m}, S_{B_1}, \ldots, S_{B_n}$ for the NetAcademies $N, A_1, \ldots, B_n$ and their vocabulary. The concrete queries $S_N, S_{A_1}, \ldots, S_{A_m}, S_{B_1}, \ldots, S_{B_n}$ are conducted in the theories of the NetAcademies $N, A_1, \ldots, B_n$ and their results are each first abstracted to the level of the Q-calculus and, then, mapped to the concrete vocabulary of N to be presented in the (concrete) language of NetAcademy N.

An agent dealing with a NetAcadamy, say N, may phrase queries using terms of vocabulary of one NetAcademy and receives results relevant to her query independent of the vocabularies used in the construction of the information being queried.

Note, that this requires the abstraction and the concretion function to be a Galois connection. Similar techniques for relating different levels of abstraction have been developed in abstract interpretation [CC78,SCK⁺95].

Further note that this mapping has only to be done for a basic subset of the vocabulary. This basic set exported via an explicit "export" construct and establishes the interface for inter-NetAcademy communication.

4   Current State of the NetAcademies

In the initial phase of the NetAcademy project, the Institute for Media and Communications Management is currently establishing three different NetAcademies covering its main areas of research. These are:

Business Media (www.businessmedia.org) deals with research on electronic markets and commerce, including reference models and pilot projects in the field of electronic data interchange (EDI) and electronic commerce.

Knowledge Media (www.knowledgemedia.org) is concerned in general innovative approaches, technologies and methodologies for knowledge management. A special emphasize is given on knowledge media, as an innovative approach for managing knowledge. Currently it deals with the representation of knowledge and information, with its semantics and inference mechanisms.

The present application areas, which serve as a base for practical feedback are in the area of strategic corporate planning, electronic commerce and scientific publishing.

Since the NetAcademy is a knowledge medium as well, the knowledge media NetAcademy provides an inside of the fundamental theoretical work for the NetAcademy itself.

Media Management (`www.mediamanagement.org`) publishes research papers and hosts discussion forums on how to analyze, define and manage the effects of new media on the economy, society, politics, law and culture.

These three NetAcademies are collected in the root of the NetAcademyNet, the `NetAcademy on NetAcademy`:

NetAcademia (`www.netacademy.org`) is the root of the NetAcademyNet and features meta-level information about the NetAcademy project and the NetAcademy platform. This NetAcademy contains the common vocabulary and general facilities like registration and administration of users.

The generic concept of a knowledge medium, which the NetAcademy concept is based on, has already been prototypically implemented as a platform for management processes. Knowledge representation, formalization, accumulation and dissemination is—like in academia—an essential part of management processes [SSRS97].

Each NetAcademy has an editorial board, constituted in analogy to editorial boards of scientific journals. Its members are responsible for the review of incoming submissions as well the control of its meta-description. According to their research direction, members of the editorial board may coordinate the review process and select reviewers.

5 Implementation

In this section, we give a brief overview of the implementation of the NetAcademy project. We begin with the user interface and its facilities to browse the NetAcademyNet and to present information. Subsequently, we will present the current state of the implementation and the technology employed to implement the NetAcademy.

5.1 User Interface

Users of a NetAcademy are unrestricted in their choice of views on the knowledge medium. When browsing, a user can dynamically and flexibly expand or narrow down depth and breadth of the information presented (c.f. Fig. 2).

The horizontal navigational bar extending across the top of the browser window allows for the selection of a particular NetAcademy to visit. Links are offered to NetAcademies closely related to the domain covered by the currently selected

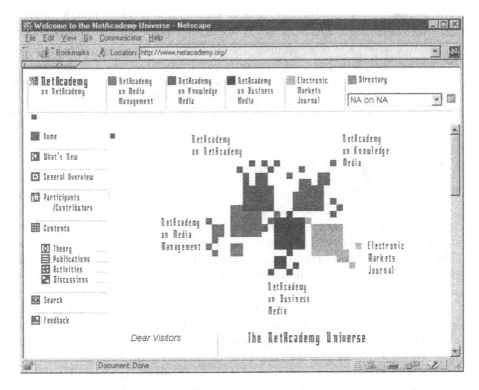

Fig. 2. Navigational interface

NetAcademy. In order to access more remote fields of knowledge the NetAcademy Global Directory needs to be consulted, which is also available via the top horizontal navigational bar.

While the horizontal navigational bar symbolizes the breadth of knowledge accessible to the user, the vertical navigational bar offers entry points into the depths of a specific domain: the individual worlds, agent- and process-related services, content, search facilities and feedback facilities.

The meta-information presented with the information to a user is classified along three attribute classes. These are:

- The *domain of knowledge* which it forms part of, i.e., the NetAcademy it belongs to.
- The *specific world* which the knowledge refers to and deals with.
- The *process or agent* which generates or transforms specific knowledge, uses it or relates and links it to other pieces of knowledge. An agent can be the author of a publication, a participant in a discussion forum but also a software agent in charge of processing queries.

Knowledge can be provided in any of the following forms and qualities: (1) *Theory*: Basic terms, definitions and axioms generally accepted and agreed upon by the scientific community of the specific domain that is covered by that particular NetAcademy. (2) *Publications* (3) *Activities*: Descriptions and presentations of research issues and ongoing projects. (4) *Discussions*: with abstracts of new knowledge, the participants agree upon.

Note, that the agents representing the entities communicating in the knowledge base have different rights to access and manipulate the knowledge stored in a NetAcademy. Each agent has a role, determining her rights to access the knowledge base.

5.2 Technology and Architecture

The technology employed to implement the NetAcademy platform has to deal with both unstructured information (e.g., papers, articles, discussions) as well as structured information (Q-vocabularies, publication attribute information and other meta-information associated with pieces of unstructured information). Therefore it was important to select components that would easily interoperate and—in combination—could accommodate structured as well as unstructured information and would allow to implement processes on top of it.

The implementation of the NetAcademy platform comprises a few pages for generic navigational structure and interface, a relational database for storing structured information and a groupware tool to deal with unstructured information and for implementing the processes.

For the storing and management of structured information a relational database management system, Oracle [Ora97], was selected. Databases were integrated for storing publications, discussions, projects and data on participants into the NetAcademy.

The system is partly based on Lotus Notes groupware technology [Lot97] for dealing with unstructured information and for implementing the processes defining the protocols for access to the knowledge. Moreover, Lotus Notes allows to enforce rules on the logical document structure as well as on layout policies and style guides. Thus, the necessary management effort to maintain the knowledge base—especially under substantial growth conditions—is reduced.

Lotus Domino technology is used for the publication of contents on the World Wide Web.

The semantic query interface is implemented as a Java applet. It offers the user the possibility to browse the vocabulary and look up vocabulary terms and construct queries from them. The interface follows a query-by-example paradigm.

Let us describe, how the vocabulary is linked to the underlying relational data model. The vocabulary is given in the Q-language. Queries to the knowledge base are phrased in the form of so called Q-tables [Kuh97,Sta97]. A Q-table for a sort is constructed from the vocabulary as a multi-dimensional coordinate system, with its attributes and their scales represented by an axis of the coordinate system. A Q-table is translated to a SQL-query, which is processed in the underlying database. The result of a query is a Q-table with entries in the coordinate system.

In order to embed a NetAcademy to a NetAcademyNet, its vocabulary has to be embedded into the common vocabulary. During embedding the specific terminology of the NetAcademy is mapped to the terminology of the common vocabulary by resolving structure differences, synonyms and homonyms. Based on the mappings the Q-calculus allows to reason about knowledge at this abstract level and to combine information from different vocabularies. Thus, we have here a federated approach [Sta97]. Each NetAcademy is independent w.r.t. information retrieval. Together, NetAcademies provide more knowledge, than the collection of single NetAcademies.

6 Related work

Today a whole range of services can be found on the Internet that claim to be or hope to become a reference source for a certain area of scientific knowledge. An example for the field of information management with focus on telecommunications is the Virtual Institute of Information [Noa96]. A theoretical foundation of the underlying concepts of knowledge organization, representation and management is not being made explicit. Furthermore, no organizational independence of the knowledge medium from the organization hosting it is attained, which tends to significantly lower the acceptance as a reference source by researchers from other institutions and might limit active contributions from users. In addition, most of those platforms are lacking sophisticated retrieval mechanisms which go beyond purely syntactic approaches.

The notion of a virtual university has gained a lot of popularity [Hei95]. In some aspects this concept is closely related to the characteristics of a NetAcademy. Just like a virtual university, a NetAcademy also aims at supporting and directing research by enabling and facilitating scientific discourse, disseminating knowledge and qualification of scientific work through reviewing processes. Unlike a virtual university, however, which like its traditional physical counterpart still represents a corporate entity, a NetAcademy stresses the goal of building a virtual reference knowledge medium independent from the institution hosting it, for a certain domain of knowledge to which a large number of researchers contribute. Moreover, the concept of the virtual university focuses on dissemination or teaching purposes using established channels, while the NetAcademy aims at renewing the process of scientific publication and management of research for the new media.

Virtual libraries provide l–much alike traditional "real" libraries– information via the Internet [Gut97,Med97]. The process of accumulation of information in such a library is not made explicit. Libraries provide information, media and search facilities based on normed data representation, but no flexible query facilities, appropriate for open, heterogeneous structure.

So called global ontologies, i.e., global schemata, are employed in [Bom93] for mediation between databases. However, the disadvantage of global ontologies is that they become large and unmanageable. Thus, solutions based on them are not scalable, in contrast to federated knowledge bases.

The Knowledge Query and Manipulation Language (KQML) [FWW+94] allows not only to relate vocabularies, but also the programs implemented on the agents. This approach is based on a global ontology and schemata as well.

Multi-lingual information retrieval mechanisms are a special case of search mechanisms with translation as a sort of mediation between heterogeneous information bases. Context information is used in [KD96,GHAW97] to increase the effectiveness of multi-lingual information retrieval. Context information is also used in [TRW+97] for search in digital catalogues.

7   Conclusions and Outlook

The NetAcademy is a medium for the accumulation, management and dissemination of information. It has the characteristics and advantages of both traditional and new media, the quality and accessibility of information and an open, distributed structure.

Still, the NetAcademy is in the early stages of implementation. The vocabulary with its query mechanisms is available as part of the NetAcademy. Currently, the process of finding appropriate editorial boards is on its way.

There are a couple of open problems due to the underlying Internet technology. The technology of the NetAcademy platform can guarantee—up to a certain point—for the authenticity of information and meta-information in a knowledge base. However, there are, up to now, no means to ensure that information and meta-information are copied and further disseminated unchanged. To make a platform like the NetAcademy useful in commercial applications copyright issues have to be resolved. Approaches to copyright issues can be found in [SL97], a concept for charging in a virtual library in [BW97].

A platform like the NetAcademy renews the process of publishing and with it the scientific work. A platform that is easy to set up and to maintain, that is able to communicate with other platforms of the same template is a new medium for researchers. We envision a growing NetAcademyNet with communicating, evolving communities organized as NetAcademies.

Acknowledgments

We are indebted to all the contributors as well as to the editors of the NetAcademies, Rolf Grütter, Axel Röpnack, Alexander Runge, Salome Schmid-Isler and Patrick Stähler.

The anonymous reviewers provided helpful comments.

Financial support for the NetAcademy project was granted by the Bertelsmann Foundation and the Heinz Nixdorf Foundation. The Swiss National Fund sponsored projects related to the NetAcademy.

References

[Bom93]   M. Boman. *A logical specification of federated information systems*. PhD thesis, Department of Computer and Systems Sciences, Stockholm University, 1993.

[BW97]   M. Breu and R. Weber. Charging for a digital library- the business model and the cost models of the MeDoc digital library. In C. Peters and C. Thanos, editors, *Research and Advanced Technology for Digital Libraries (ECDL'97)*, Lecture Notes in Computer Science 1324, pages 375–386. Springer-Verlag, 1997.

[CC78]   P. Cousot and R. Cousot. Static determination of dynamic properties of recursive procedures. In E.J. Neuhold, editor, *Proc. 2nd IFIP TC-2 Working Conf. on Formal Description of Programming Concepts*, pages 237–277. North-Holland, August 1978.

[FWW⁺94] T. Finin, J. Weber, G. Wiederhold, M. Genesereth, R. Fritzson, D. McKay, J. McGuire, R. Pelavin, S. Shapiro, and C. Beck. Specification of the KQML agent-communication language, 1994. URL: http://logic.stanford.edu/papers/kqml.ps.

[GHAW97] R. Gaizauskas, K. Humphreys, S. Azzam, and Yorick Wilks. Concepticons vs. lexicons: An architecture for multilingual information extraction. In M.T. Pazienza, editor, *Information Extraction – A Multidisciplinary Approach to an Emerging Information Technology*, Lecture Notes in Artificial Intelligence 1299, pages 28–43. Springer-Verlag, 1997.

[Giu97]   B. Giussani. A new media tells different stories, 1997. Zielgruppe unbekannt— Wer nutzt das Internet?, Berner Technopark.

[Gut97]   Project Gutenberg. Project Gutenberg - fine literature digitally republished, 1997. Available at: promo.net/pg/.

[Hei95]   H. Heilmann. Editorial: Virtuelle Organisation. In *Handwrterbuch der Modernen Datenverarbeitung: Virtuelle Organisation*, volume 185. 1995. 32. Jahrgang.

[KD96]   A. Kosmynin and I. Davidson. Using background contextual knowledge for documents representation. In C. Nicholas and D. Wood, editors, *Principles of Document Processing (PODP'96)*, Lecture Notes in Computer Science 1293, pages 123–132. Springer-Verlag, 1996.

[Kuh97]   C. Kuhn. *Designing a market for quantitative information*. PhD thesis, Institute for Information Management, University of St. Gallen, 1997.

[Lot97]   Lotus Development Cooperation. *Lotus*, 1997. See: www.lotus.com.

[Med97]   Medoc. M$_E$DOC-the online computer science library, 1997. Available at: medoc.informatik.tu-muenchen.de.

[Noa96]   E. Noam. What is the v.i.i.: About the institute., 1996. URL: http://www.ctr.columbia.edu/vii/mwhat.html.

[Ora97]   Oracle Corporation. *Oracle*, 1997. See: www.oracle.com.

[Sch97]   B. Schmid. The concept of a NetAcademy. Institute for Information Management, University of St. Gallen, 1997.

[Sch98]   B. Schmid. *Wissensmedien*. Gabler-Verlag, 1998. To appear.

[SCK⁺95] B. Steffen, A. Claßen, M. Klein, J. Knoop, and T. Margaria. The fixpoint analysis machine. In I. Lee and S.A. Smolka, editors, *6th Int. Conf. on Concurrency Theory (CONCUR'95)*, Lecture Notes in Computer Science 962, pages 72–87. Springer-Verlag, 1995.

[SGW+96] B. Schmid, G. Geyer, W. Wolff, R. Schmid, and K. Stanoevska-Slabeva. Representation and automatic evaluation of empirical, especially quantitative knowledge, March 1996. Final report of the Swiss National Science Foundation Project No. 5003-034372,.

[SL97]   M. Stefik and G. Lavendel. Libraries and digial property rights. In Carol Peters and Constantino Thanos, editors, *Research and Advanced Technology for Digital Libraries (ECDL'97)*, Lecture Notes in Computer Science 1324, pages 1–10. Springer-Verlag, 1997.

[SSRS97] B. Schmid, T. Schwan, A. Röpnack, and M. Schwartz. Enterprise Knowledge Medium (EKM): Konzeption eines Mediums zur Unterstützung von Führungsprozessen. *DV Management*, 2, 1997.

[Sta97]   K. Stanoevska. *Neugestaltung der Unternehmensplanung mit Hilfe eines prozessorientierten Planungsinformationssystems*. PhD thesis, University of St. Gallen, 1997.

[TRW+97] K. Tochtermann, W.-F. Riekert, G. Wiest, J. Seggelke, and B. Mohaupt-Jahr. Using semantic, geogrphical and temporal relationships to enhance search and retrieval in digital catalogs. In C. Peters and C. Thanos, editors, *Research and Advanced Technology for Digital Libraries (ECDL'97)*, Lecture Notes in Computer Science 1324, pages 73–86. Springer-Verlag, 1997.

Distributed Compression of Live Video –
An Application for Active Networks

Robert Hess, Dagmar Geske, Sascha Kuemmel, Henrik Thuermer

TU Dresden
Centre for Highspeed Networks and Multimedia
Multimedia Application Group
hess|geske|kuemmel|thuermer@ibdr.inf.tu-dresden.de

Abstract. A rising number of video conferencing applications uses software codecs for video compression. This allows to support several compression schemes without the necessity for expensive dedicated hardware for every new standard. Using inexpensive framegrabber boards, it is possible to support several different compression schemes as H.261 or H.263. But the computing power of today's desktop system does usually not allow for the simultaneous compression and decompression of high quality video. One solution is to equip these systems with dedicated hardware, trading off the above mentioned flexibility and introducing considerable costs per machine. This paper presents an approach for the distributed compression of live video, exploiting cheap LAN capacity and thus dividing the compression process into a short pre-processing on the desktop and a second stage performing the more complex operations on a dedicated compression server, that may be shared by a workgroup or even more people.

1 Introduction

Main focus of this paper is the distribution of the coding process of common, DCT-based video compression schemes with inter-picture dependencies. Transport-level issues are of great importance for this approach and are part of ongoing efforts or are described elsewhere [1].

1.1 Problems

Since software video codecs are employed more and more for commercial conferencing systems, some new questions arise with respect to the resources of end systems and regarding performance issues.

On a common desktop computer the user deploying software compression has to experience either 1.) a low quality video or 2.) a very high workload that hinders the efficient use of other programs, i.e. application sharing, which may be necessary in a conferencing situation. On low end machines, the user may experience both kinds of limitations.

Hence there is the demand to lighten the burden of the desktop system and to improve video quality with the same amount of processing time.

The common approach to solve this demand is to employ dedicated hardware for the compression of the video stream. This approach, however, has several drawbacks. Common hardware does usually support only one coding scheme, especially support for newly emerging standards can not be given. Additionally costs per system are introduced, these may form a considerably part of the systems overall costs. For these reasons the hardware approach does not scale well for a large number of systems and for general changes or additions in the compression schemes supported. This counts especially, if a company has to support multiple standards with different coding schemes because of the different partners they have to communicate with.

1.2 Solutions

Analyzing these problems, we come up with two goals: 1.) solving the problem of insufficient performance on desktop systems with a solution, that scales to new standards and to large numbers of clients. 2.) limiting the costs per system for introduction respectively changing of a video compression facility.

One way to reach these goals would be to improve compression algorithms in a way that they perform much faster. Of course there is a limit for the speedup that can be reached there. This is given by the nature of compression that relies on redundancy and on the omission of insignificant information. Especially finding temporal redundancy requires searching and the decision which information is insignificant requires transformation of the picture. That means that both, searching and transforming do not scale well, and rise with image size and that for a given image size and quality there is a lower limit in processing time we can not beat.

We will show the results and limitations of these approach in the following chapter, but unfortunately the performance of today's desktop systems allows only for a rather poor quality. Since one can expect that with rising computation power also the users expectation concerning size and quality of the video will rise, one can not hope to solve the problem by waiting, at least not in the next few years.

Because of this situation, we propose another solution, were we can exploit the fact, that not all members of a workgroup will be in a videoconference simultaneously (except in the case of video call centers or similar facilities) and the fact that in the workgroup area, LAN-capacity is a comparably cheap resource today. So we come up with the splitting of the compression process into a first stage, that will be performed on the desktops, a pre-compressed image that will be transmitted over the LAN and a second stage that will complete the compression procedure. This steps will be described in detail below. Furthermore it will be shown, that these approach solves our problems of scalability and costs in a certain degree and that it is suitable to a whole class of DCT-based compression schemes relying on inter-frame dependencies. Such a kind of application maps well into the *active network* concept [2] and is related to concepts like video gateways (transcoders)[3]. This aspects will be discussed in some detail in chapter 4.

The paper consist of two main parts. In the first part some explanations about a H.261 and a H.263 codec developed by our group are given. These codecs are the base for our performance figures. The second part focuses on the aspect of distribution of the compression process.

2 Compression Process

Our evaluation is based on two codecs we have implemented and tuned especially for the investigation of live compression for high quality video on the Intel plattform under Windows NT 4.0. We've chosen the H.261- and a H.263 compression schemes since they are widely employed an standardized by the ITU-T [4], [5]. Both of our codecs are based on the idea to accept larger output streams to achieve a significantly reduced compression time. This allows for real time execution with CIF images and 25 fps in a very good quality on high-end machines. Nevertheless this process consumes nearly all of the computing power. On average desktop systems only a very low frame rate with a poor quality can be achieved. A detailed description of the H.261 codec can be found in [6].

2.1 Main Stages of Compression

In the following a very short description of the compression process in H.261 and H.263 is given. Since this is a wide field, the reader may consider to read [7] for an introduction or [4], [5] for the details. Fig. 1. gives an overview over the single steps. The compression process consists mainly of the following steps:

- The image is divided into macroblocks which consist of several blocks containing either luminance or chrominance information.
- These macroblocks can be code in intra or inter mode. Intra mode uses the original image, inter mode some kind of difference between the last and the current picture. Instead of the last picture, usually the compressor (coder) reconstructs the image from the compressed data to have the same input as the decompressor (decoder).
- For macroblocks transmitted in inter mode, a motion estimation may be applied, that means a searching for the macroblocks position in the previous picture. This allows to transmit only a motion vector instead of a macroblock.
- In the next step, the remaining macroblocks were transformed using discrete cosine transform (DCT) to prepare data for discarding in quantization.
- The quantization process sets all data to zero, that contains no information important to the human eye. The degree of quantization mainly corresponds with resulting image quality - higher quantizers mean lower quality.
- The quantized coefficients are compressed using entropy coding.

At the receivers side, the decompressor has to perform the inverse steps. That means:

- Decompress the entropy coding.
- Apply inverse quantization.
- Apply inverse DCT.
- reconstruct blocks given by motion vectors, if it is inter coding.
- Add the result to the corresponding macroblocks from previous picture, if it is inter coded, in intra mode add whole macroblock to image.

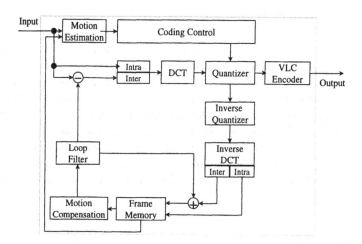

Fig. 1. Steps of the coding process in a sample H.261 coder

2.2 Assumptions

The design goal for these codecs was to make as many assumptions as possible to speed up the common case.

- Highly specialized codec for the conferencing case, that means half-length portrait and very moderate motion. This allows to apply very specific motion estimation schemes.
- Highly integrated processing of all different stages from capturing/compression and decompression/displaying. Hence ideas from concepts like integrated layer processing (ILP) [8] and application layer framing (ALF) [9], [10] should be applied, where appropriate. It should be possible to improve speed significantly due to reduced copy operations.
- The quality aimed at should be CIF (352x288) with approximately 25 fps. For future projects we hope to achieve even larger images (exploiting mainly hardware speedup and better OS-Support).

We're aware that 15 fps are said to be sufficient, but had to learn, that such a claim does not satisfy high tech professionals wanting to collaborate or bank staff talking to customers.

- With respect to network transportation, the size of the handled data units and possibilities for fast and simple error correction schemes should be evaluated. This approach allows to keep latencies low, since no retransmission is required in case of minor errors. Instead data from previous frames may be used.
- We considered primarily LAN and Broadband WAN connections, so the size of the resulting data streams is not of primary concern. This assumption seems necessary for compression on a desktop, since today's processor speed does not allow extensive motion vector searching, but it is also common in other systems like MMC and NetMeeting when gaining for good video quality. This assumption is not necessary with distributed compression as shown later. Lower bandwidth can generally be achieved without changes by lowering the frame rate.

In the following we will focus only on a description of the H.261 implementation since all questions related to distributed video compression are similar with respect to H.263.

There are several public H.261 implementations available. We investigated three of them, these were the p64 of PVRG [11] and the H.261-codecs implemented in the mbone tool VIC [12] and in the INRIA video conferencing system IVS [13], [14]. These implementations each focus on different aspects of the H.261-standard, especially concerning the time critical encoding process. The H.263 codec based mainly on the TMN-codec. [15]

2.3 Implementation

For the implementation, two approaches were chosen following the results of other implementations:

For fast compression, ideas derived from the VIC-implementation were used. This implementation will be referenced in the following as the FAST implementation.

Alternatively a fully featured H.261 coder for achieving lower output rates by means of more sophisticated algorithms (and thus slower operation) was realized. This case will be referenced as the FULL implementation.

FAST. For the first approach, which can be described as the art of not to do unnecessary things, we went a step further in using fix point operations for the DCT and an optimized algorithm for adaptation of smaller quantization step sizes. Thus the computing time could be reduced by approx. 15 to 25 % .

These optimizations did taint neither quality nor output rate. These results confirm basically with those derived in [16], [17].

FULL. For the second approach the structure of the VIC-coder was enhanced by a section for motion estimation, an algorithm for determining the type of macroblocks (as in p64), and sections for the coding of INTER-frames. The reconstruction of the coded frame was derived from the VIC-decoder.

For motion estimation two algorithms are used for different quality requirements which perform at comparably lower costs than full search as realized in p64. The first is a search in a diamond area and the second a 2-dimensional logarithmic search [7]. These algorithms were chosen as result of extensive experiments with 5 different search algorithms.

Further speed up was reached by using motion detection as in the VIC-coder prior to motion estimation. So all macroblocks that were identified as „not moved at all" could be excluded from costly motion estimation.

Another point that can be used to reduce computations is, that after DCT and quantization a number of INTER-macroblocks are discarded due to their low content of information. It can be seen, that all these blocks have a *mean square error* (MSE) significantly different from the rest of macroblocks. Since MSE is needed by the algorithm for determining the type of macroblocks, it is cheap to use this feature for discarding these blocks before DCT is performed. Despite the comparably high cut down in computing time, quality and efficiency of compression suffer only little.

Comparing the two modes we've chosen to implement, it is to be seen, that, assuming the same quality (signal to noise ratio - SNR), the FAST implementation only needs approximately 20-30 % of the computation time the FULL-implementation consumes. Nevertheless the later results in an output stream of only half the size of the FAST-version's. These figures illustrate, that nowadays, given sufficient bandwidth, only the FAST approach is satisfying for software compression on desktop machines. This may change in a couple of years. It has to be stated here, that for instance Microsoft uses a kind of the FULL approach in its codec for its netmeeting, since in the typical semi-professional environments for these conferencing system bandwidth is of a higher concern than video quality.

2.4 Performance Evaluation Methods

Evaluating the performance of a content-dependent video compression scheme is a difficult task. There are a lot of dependencies and it is nearly impossible to give exact figures for others to compare with. Besides the obvious parameters as frame rate and image size, performance data depends also on:
- Contents of the compressed video.
- Implementation of color space conversion (rgb -> yuv), if this step is necessary.
- Compiler, its options, and sometimes, even with high optimizing compilers[1], the Order modules are arranged within code.
- Even within the same machine class, there may be subtle differences i. e. due to the size of processor cache.

---

[1] It might be mentioned, that manual optimization of C-Code often did not lead to the expected results. Sometimes performance changed stochastically and often we had to learn from inspecting assembly code, that the compiler itself had always done the job without human interception. So it has to be stated, that manual optimization seems to be of minor importance with highly optimizing compilers as the used MS Visual C 4.2 / 5.0 compilers are. Nevertheless it is always important for maximal performance gains to have a close look to compiler output at crucial points (inner loops etc.).

The quality of the resulting video is also an issue for further discussion. Common measures as SNR (Signal to Noise Ratio) or PSNR (Peak SNR) are only meaningful comparing the same clip compressed with similar methods. Its not very useful for comparing different video clips or totally different compression approaches, since the kind of compression artifacts may differ with respect to contents and algorithm. To get an impression of the quality a human spectator would experience, the only way is to have several people watch the sequence, compare it to other sequences of known quality and to give their impression to build a mean opinion score (MOS). This is however not a very handy approach for daily lab practice.

To give nevertheless meaningful performance figures we've chosen to compare our codec with other codecs available to us and to use identical machines and identical video clips. So we can assume that SNR-Values represent quality differences in a sufficient way.

The operating system used was NT 4.0 and the codec was integrated into the Video for Windows architecture. The MS Visual C Compiler Version 4.2 / 5.0 was used.

2.5 Performance Comparison

Compared with commercial codecs as the msh261.drv (Microsoft) and the dech261.drv (Digital) which ship with Microsoft NetMeeting our approach manifests the following facts:

- Given the same SNR, our FAST approach performs significantly faster resulting in higher output streams.

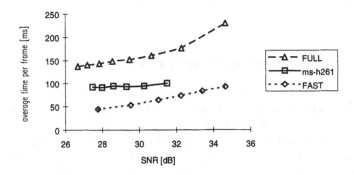

Fig. 2. Performance of our FAST and FULL-implementation of the coder compared to Microsofts coder from netmeeting. Data was taken on a 133 MHz Intel Pentium

- Our full approach is despite our efforts to reach an highly optimized implementation slower than the Microsoft product, but somewhat faster than the one from Digital.
- Comparing computing time and the achieved output rate for coding and decoding at different SNR-levels of our codec with the H.261 codec from Microsoft shows,

that their coder is highly optimized and achieves output rates from 20 to 30% below the rates of our fully featured coder with a lower computation time. In this area the Microsoft coder is obviously superior. The main application field of our codec are but high frame rates with a high quality „regardless" of output rates. As this was the aim of our design, we reach with our FAST implementation in this area remarkably better results (40 to 60 %).

- The H.261 Codec from Digital Equipment Corp. delivered with Microsoft NetMeeting for Alpha, shows some flaws in reading its own streams. Besides the use of the input format YUY2 causes a faulty stream. In terms of output rate it is nearly as good as our FULL-codec, the computation time is 10 to 15 % higher.

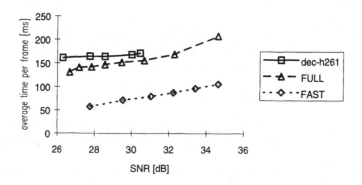

Fig. 3. Performance of our FAST and FULL-Implementation compared to Digitals Coder from netmeeting. Data was taken on a 233 MHz Alphastation 400.

2.6 Detailed Performance Analysis of the Process

The following table shows the average computation times in ms for the single steps of both H.261 implementations at an rather good SNR-level (32,3) on a Pentium 133 with 512 KB Cache and 64 MB RAM.

These steps are accompanied with a reduction of the input stream. Since several steps lead to different degrees of reduction, it is very important to find the right step for dividing the process into two parts for distribution. Obviously the FULL implementation is not the first choice for distribution, since the conversion to YUV 411 only leads to a maximum reduction of a half (in case of RGB 32) and Motion Compensation is much to costly. So only the FAST case seems to be suitable for distribution.

The overall process consists of a *motion detection* (different to *motion estimation*), discrete cosine transform with a following quantization and a run length coding of the quantized values.

Tab. 1. Times for the compression stages of the FAST and FULL-Implementation of our H.261 Codec

FAST	ms	FULL	ms
YUY2-> planar 411 YUV	6	YUY2 -> planar 411 YUV	6
motion detection	2	motion estimation	66
DCT + quantization	22	DCT + quantization	48
adaptation	5	adaptation	0
Huffman	11	Huffman	8
		GetCBP	7
		loop filter	12
total	46	total	145

Values from our H.263-codec show a slightly different behavior. The results are for a SNR-Value of 27,1 dB and a PSNR of 36 dB. (For comparison with the H.261 values see remarks about performance evaluation methods).

Tab. 2. Times for the compression stages for the FAST and FULL-Implementation of our H.263 Codec

FAST	ms	FULL	ms
YUY2	6	YUY2	6
motion detection	3	motion detection	6
		motion estimation	10
		halfpixel prediction	12
DCT + quantization	18	DCT + quantization	20
VLC + Huffman	2	VLC + Huffman	4
total	25	total	63

Since we have realized a combination of motion detection and motion estimation, that means motion estimation is performed only with the macroblocks that have changed, we can divide also the FULL implementation after motion detection. This is especially interesting, since the FULL approach results in significantly smaller streams (approx. half of the fast approach).

This combination of motion detection and motion estimation is, by the way, not a feature distinct to H.263, indeed we have first tested it for H.261. But due to lack of resources we have only optimized the H.263 version by now. More interesting features special for H.263 are for instance halfpixel prediction, leading to a visible higher quality with same computation time.

3 Distribution

Now we will explain in detail how the distributed compression process works. As an example we will refer to a Pentium 133 with 512 KB Cache and 64 MB RAM as low end desktop system and a P II 233 MHz with 128 MB RAM as high-end desktop system. A PentiumPro 200 with 256 KB Cache and 128 MB is the compression server for H.261 and the above P II 233 for H.263.

The cycle time per frame for a video with 25 fps is 40 ms and for 15 fps 66 ms, that means all non-compression operations as framegrabbing, network handling and displaying have also to fit into this time slot. Since the video should have CIF-size, an input stream of approx. 30 Mbit/s for 25 fps and 18 Mbit/s for 15 fps is generated using a 12 bpp-format. For the YUV9 format we have 22 Mbit/s and 13 Mbit/s accordingly.

3.1 Compression

First Stage. The first stage includes a *motion detection* as described above. Here we need only approx. 5% of the overall time. Another time is added for the conversion of the input format, usually framegrabbers support some kind of YUV-format that has only to be subsampled and arranged into the fitting planar format. In the previous chapter we introduced 6 ms for a conversion from YUY2. The availability of a planar format with an optimal subsampling as for instance yuv9 would further reduce this time.

The size of the resulting stream depends on the threshold used in motion detection. Mean reductions from 30 to 50 % of the original size could be reached with good quality. This would mean a stream of 4,2 to 15 Mbit/s for LAN-transmission from desktop to compression server. This would be a reasonable value given a usual 100 Mbit/s shared media LAN. The size of this stream depends mainly on the algorithms used. For higher capacity LAN or switched solutions on the one hand, and slower LAN's as switched 10 Mbit/s on the other, several adaptations of the input stream and the threshold value are possible to achieve optimal performance and load balance.

Second Stage. The second phase transforms the incoming stream into DCT-Coefficients and quantizes them. The final phase is the preparation for the entropy encoding and the encoding itself. The resulting stream depends on the used algorithms an is between 170 and 700 kbit/s in our examples. For real time operation the server needs to receive, process and send the frame in a maximum of the time slot at the given framerate (40 ms for 25 fps, 66 ms for 15 fps). To be faster would not help in anything then reducing delay introduced by the distribution. This delay has a (theoretical) minimum of one frame.

Scalability, Adaptability. There are some interesting points to discuss at this point. As shown in our experiments with the FULL codec, it is possible, to introduce a motion estimation in second stage after motion detection in first stage thus reducing further the size of the output stream. Another point is, that the decision, which coding standard we apply is taken not in the first, but in the second stage. That means a compression gateway could communicate with heterogeneous partners transparent for endsystems and solves thus the problem of „simply" upgrading to emerging standards etc. by only adding software to the central server. The problem of scalability in terms of the number of users can be solved by extending the computing power of the server or simply by adding servers, that are dynamically selected.

3.2 Decompression

Given sufficient LAN-capacity there is the option to place nearly the whole decompression process on a central server. This leaves a larger amount of free CPU-capacity on client machines for other tasks. While this is desirable in general, it has to be mentioned, that it is also possible in some cases to perform decompression on the client side to save LAN-capacity.

First Stage. The first stage includes the full decompression process to the point of fully reconstructed macroblocks. These are transmitted to the client system. The resulting stream is a little bit smaller than the stream between client and server in the compression process. Few macroblocks may be discarded due to motion estimation, but the preceding motion detection limits the number of such macroblocks.

Second Stage. The second phase places the reconstructed (and thus changed) macroblocks at the corresponding positions into the previous frame.

This partitioning leaves only minimal computation to the client but requires the transmission of a much larger stream from server to client.

3.3 Overall scenario

The following table shows complete scenarios for a duplex video transmission as common in conferencing. These values are mean values and depend of course on the video transmitted. All scenarios are conservatively configured to leave sufficient extra network and cpu capacity for burst transmission and other applications.

H261. The FAST implementation was used with a quantizer of 5, a threshold for motion detection of 48 at 15 fps. That results in an average transmission of 30% and a maximum of 40 % of all macroblocks. The resulting average stream is only 4,2 Mbit/s since a YVU9 format with 9 bits per pixel is used as input. This stream has to be transported in both directions, since no further motion estimation on server side is applied and no further blocks can be discarded. The compression ration however is not excellent with the used intra-coding, effective compression from 9 bit input format is only approx. 1:20 and the resulting 700 kbit/s are higher then today usually affordable for an office scenario. LAN load with an average of 8 Mbit/s may be affordable with 100 Mbit/s capacity.

It would be possible to perfom decompression completely on client side, this would reduce net-transfer to 5 Mbit/s but increase average CPU-load on the client machine to approx. 80 %. This would mean to block the machine for every other task.

CPU load on server would allow for another compression session to run.

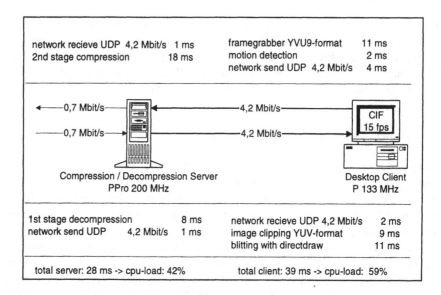

Fig. 4. Scenario for distributed compression using FAST H.261 with distributed decoder and 15 fps

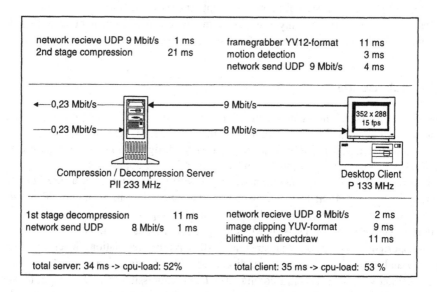

Fig. 5. Scenario for distributed compression using FULL H.263 with distributed decoder and 15 fps

H263. For the H.263 codec two scenarios with inter coding (FULL) were shown. They allow a significantly higher compression ratio from 1:50 to 1:100. For this example the compression rate of 1:100 was chosen, since the higher compression leads to shorter compression and decompression times due to the smaller amount of data to be processed.

The scenario with 15 fps results in an average load of 50 % in both, client and server. Here we have also a very low output rate with approx. 230 kbit/s. This would be optimal for a conferencing over 3 ISDN lines. LAN load however is rather high. This underlines the better suitability of the 9 bit input format. Of course quality loss due to the reduction of the input format does not reflect in the SNR-value and hence is not measured here.

The 25 fps-scenario comes with even higher LAN loads but only slightly higher rates after compression. Here we need a high-end desktop system to fulfill our requirements. These results show that a very good quality requires a remarkable amount of resources in terms of LAN and WAN bandwidth and of computing power.

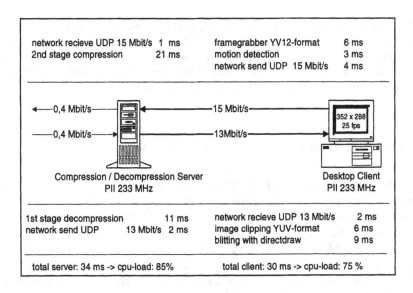

Fig. 6. Scenario for distributed compression using FULL H.263 with distributed decoder and 25 fps

3.4 The Benefits of Distribution

The results shown above indicate, that distribution of the compression process clearly help to improve performance of video compression at the desktop. The following figure shows the average load for all of the three possible scenarios. It is to be seen that especially the distribution of both, the coder and the decoder reduces the load significantly. The tradeoff however is to replace compression with network transport. Especially in slow or heavy loaded LANs this may consume all the benefits of distribution. In our example with a fast 100 Mbit/s LAN with medium load conditions it is however a very good ratio between network transmission and compression time.

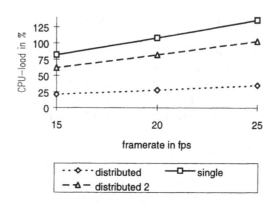

Fig. 7. Client performance gains for different distribution models, the classical model (*single*), the distribution of coder and decoder (*distributed*) and the distribution of only coder (*distributed 2*). This Data is for a H.263 inter-coding (FULL) with approx. 27 dB SNR at a 133 MHz Pentium. These values reflect only compression / decompression and network transport, since capturing and displaying have constant times for all three models

4 Application in an (active) network scenario

The motivation for introducing distributed compression is similar to a company telephone system (or corporate network) based on customer owned PBXs. Nobody tries to configure such a system with a count of „lines" between customer and service provider equal to the count of telephones of the company. We assume that the ratio of installed video conferencing client systems to simultaneously used external video connections is similar to the ratio of telephones to external telephone lines. Furthermore we should consider conferences with more than two participants. In this case we can have two or more users inside the company (e.g. one sales representative and two member of the technical staff) and only one user (e.g. customer) outside. Fig. 8 depicts the information flow in such a case. This shows that, depending on the application, the count of internal connections may be significant higher than the of externals. Internal connections could be realized without involving compression servers and depend only on the available LAN resources.

4.1 Universal Compression Service

A universal video compression service for workgroups or larger groups connected via high-speed LAN connections. uses a central pool of compression resources. Therefor a dynamical assigning of compression servers to clients depending on server workload is necessary. Here traditional approaches for distributed resource allocation as used i.e. in Corba could be applied.

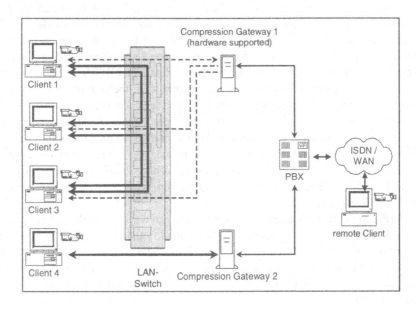

Fig. 8. Complex scenario for the application of compression servers in an enterprise

As mentioned above, such a distributed compression system can be designed in a manner, that the client part performs only generic operations for compression/decompression, that are not specific to a special compression scheme. Of course minimal assumptions must be taken, so the compression algorithm has to work with the same division into blocks, use the same subsampling for Luminance and Chrominance and it has to rely on interframe dependencies.

If these facts are given, it is possible to decide on server side, which compression scheme to use. This works for both compression and decompression. Hence all effort to introduce a new compression scheme is to update server software.

In Fig. 8 a complex scenario for such a compression service is depicted. It will also be possible and meaningful to use hardware based codecs instead of software compression in the proposed distributed compression system. Such compression hardware should be established in the second stage to provide support for more clients simultaneously or to achieve a better video quality. In this scenario the overall costs will be not much higher than for software based video compression, because hardware support is only necessary in server systems. To preserve the advantage of flexibility the combined use of software and hardware based codecs is recommended. The support of a new compression standard requires thus only the installation of a new software based gateway. If the number of clients using the new standard simultaneously rises, the software based server could be substituted by a hardware accelerated system. However, no changes on the client systems are required.

4.2 Application Gateways (Transcoder)

Since there are different standards concerning videoconferencing in different environments (e.g. line base vs. IP-based, narrowband vs. broadband) there is an inherent interworking problem between these systems. Several standards (i.e. H.321) assume the existence of application gateways, that bridge between networks with different characteristics and thus handle conversion of video formats. Such transcoders were also mentioned in [2] and described in detail in [3]. The scenario described above may be easily applied to this case. Closely seen, the compression server in our example is always an application gateway, that bridges between a broadband LAN and a comparably narrow WAN and thus allows the precompressed streams, that might be easily used for inhouse conferencing without alteration, to pass to another application domain. Given a broadband access to the customer, the compression server could be located at the network providers facilities too. This requires of course an adequate tarifing model, selling the whole service at a competitive price instead of seeing only the bandwidth consumed.

Furthermore the principle may be applied for gateways inside the provider network, that decompress streams to a necessary level, modify the information, and compress it again in another compression scheme. This would consume significantly fewer resources than a total decompression and new compression and would save a great part of the videos quality since the lossy part of the process would be applied only once for the stream.

5 Outlook and future work

The above described activities are part of larger efforts, which include also the questions of network transportation, user interface and the mapping of transport-QoS to user visible video quality. Here we are currently on the way of implementing and testing a complete toolkit for video transmission, which allows also the combination of different streams in order to investigate the effects of multimedia streams in complex networks. Our transmission system features also a newly developed transport protocol, that allows to send and receive streams with a considerably lower CPU-load than UDP does (not to mention TCP, that has even more overhead). Details regarding these protocol are mentioned in [1]. Using this protocol we can further decrease the cost for communication between client and server.

The transmission toolkit has an error detection scheme, that uses the data from the previous frame instead of re-transmitting the data.

In order to reduce local LAN load, we will implement and evaluate further possibilities to divide the compression process. This will allow to fine tune the distribution of components depending on LAN-capacity and computing power of client and server. Furthermore a system for the dynamical distribution of the single steps of process is planned.

6 Summary

An approach for distributing the compression of live video is shown by the example of H.261 and H.263 coding. This approach should work for the majority of coding schemes relying on temporal redundancy.

It was shown that our approach might help 1.) solving the problem of insufficient performance on desktop systems with a solution, that scales to new standards and to large numbers of clients. 2.) limiting the costs per system for introduction respectively changing of a video compression facility.

References

[1] S. Kümmel, T. Hutschenreuther. Protocol Support for Optimized, Context-Sensitive Request/Response Communication over Connection Oriented Networks. IFIP Int. Conf. on Open Distributed Processing, Toronto, May 1997.

[2] D. Tennenhouse, J. Smith, W. Sincoskie, D. Wetherall, G. Minden.. A Survey of Active Network Research. IEEE Communications Magazine, Vol. 35, No. 1, pp80-86. Jan. 1997.

[3] E. Amir, St. McCanne, and H. Zhang. An Application-level Video Gateway. ACM Multimedia '95, San Francisco, CA, November 1995.

[4] ITU-T Recommendation H.261. Line Transmission of non-telephone signals.Video codec for audiovisual services at p x 64 kbit/s, March 1993.

[5] ITU-T Recommendation H.263. Line Transmisson of non-telephone signals. Video codec for low bitrate communication

[6] D. Geske, R.Hess, S. Kuemmel. Fast and predictable video compression in software - design and implementation of an H.261 codec, to appear in: Interactive Multimedia Service and Equipment - Syben 98

[7] V. Bhaskaran and K. Konstantinides. Image and Video Compression Standards, Algorithms and Architectures. Kluwer Academic Publishers, Boston, 1995.

[8] T. Braun, C. Diot. Protocol Implementation using Integrated Layer Processing. SIGCOMM 1995.

[9] B. Ahlgren, P. Gunningberg, K. Moldeklev. Performance with a Minimal-Copy Data Path supporting ILP and ALF. 1995.

[10] S. Floyd, V. Jacobson, C.-G. Liu, St. McCanne, and L. Zhang. A Reliable Multicast Framework for Light-weight Sessions and Application Level Framing. IEEE/ACM Transactions on Networking, November 1996.

[11] A. C. Hung. Understanding Image Compression. PVRG, Stanford University, 1993.

[12] St. McCanne and V. Jacobson. vic: A Flexible Framework for Packet Video. ACM Multimedia '95, San Francisco, CA, November 1995.

[13] T. Turletti. H.261 software codec for videoconferencing over the Internet. INRIA Research Report Journal no 1834, January 1993.

[14] T. Turletti. The INRIA Videoconferencing System (IVS). ConneXions - The Interoperability Report Journal, Vol. 8, No 10, pp. 20-24, October 1994.

[15] detailed information and sources may be found here: *http://www.nta.no/brukere/DVC/*

[16] P. Bahl, P. S. Gauthier, and R. A. Ulichney . Software-only Compression, Rendering, and Playback of Digital Video. Digital Technical Journal, April 1996.

[17] D. Simpson, A. Swan, R. Thomas. Architectural Influences of DCT Based Software Video Decoders. May 1996.

Incremental Scene Graph Distribution Method for Distributed Virtual Environments

Ken'ichi KAKIZAKI

Department of Computer Science and Electronics
Kyushu Institute of Technology
680-4 Kawazu, Iizuka, Fukuoka, 820, Japan.
kakizaki@cse.kyutech.ac.jp

Abstract. This paper describes an incremental scene graph distribution method for client-server based distributed virtual environments. It is very important for distributed virtual environments that all users share the same environment on a service. In order for users to share an environment, a scene graph that describes the virtual environment must be distributed to all clients by the server. The method we propose divides a scene graph into segments and distributes these segments incrementally. The distribution is performed according to the importance of the object's visibility, and the measurement of importance is evaluated by each client. This measurement is performed in a traversing process for scene rendering, so it does not expend any extra processing cost. If an important but still as yet undistributed scene graph segment is found by a client, the client requests its distribution of the server, at which time the server distributes the segment upon demand. The method realizes an efficient scene graph distribution in a narrow bandwidth network.

1 Introduction

In recent years, implementation methods of distributed virtual environments have been well researched, and the technology is expected to become an excellent service platform in the network [Maxfield95, Broll95, Stansfield95]. Many researchers have focused on a data exchange method for dynamic information such as event data and motion data [Barrus96, Kessler96, Macedonia95, Singh95]. Efficient dynamic information exchange methods are very important: However, we believe that a scene graph distribution method is equally important for distributed virtual environments.

It is very important in distributed virtual environments that all users share the same environment on a service. In order to share the same environment, a scene graph that realizes the virtual environment must have been received from somewhere else previously. However, almost all researchers assume that the use of the distributed virtual environment will be used by the people who work together, and that the network bandwidth would be as wide as a local area network. They also assume, therefore, that a scene graph has been previously distributed to each user's machine. However, if a distributed virtual environment is to be used by an unspecified number of people, as is the case with the WWW (World Wide Web), it would be difficult to have had this scene graph distributed previously. In this case, a user must download the scene graph at the start time of a session.

Downloading is a well-known method amongst VRML (Virtual Reality Modeling Language) users, and those users also know that the download time is long, especially with a large scene graph. Users must wait until the scene graph is completely downloaded, and users cannot actually see the visual scene of the virtual environment until the download is completed. This situation can clearly be shown in the case of a narrow bandwidth connection, especially in modem based connections, and when the amount of scene graph to be distributed is large. For example, the download time for a scene graph that the amount is 10M bytes is at least one hour. This situation can lead to heavy stress for users. In order to accommodate a widely accessible distributed virtual environment like the WWW, we must introduce an efficient scene graph distribution method.

This paper describes an incremental scene graph distribution method for distributed virtual environments. A scene graph for a large virtual environment is constructed from a large amount of data. However, a user can see only a small part of the environment at one time, so reference for the scene graph has locality. Therefore, the user only needs a small part of the scene graph which describes near his or her viewpoint instead of the whole of the scene graph data. We introduce a scene graph distribution method that uses the locality effectively. In order to apply the locality, our method evaluates the visible importance of objects, and distributes a scene graph incrementally, according to the object's importance.

2 Distributed Virtual Environment

In this chapter, we show the system overview we assume; we discuss a scene graph distribution method for this system.

2.1 Client Server Model

The distributed virtual environment system is constructed based on a client server model. The relationship between clients and servers is shown in **Figure 1**. In this system, every client can connect to all servers. This concept of connection is similar to that of the WWW. Many clients can connect to one server at a time, and the connection is used to exchange information in order to provide the distributed virtual environment service. The flow of information exchange is shown in **Figure 2**.

Each server which provides an individual distributed virtual environment service to clients, is connected to the server. In order to provide a service, a server distributes a scene graph that describes its virtual environment, gets event information from clients, and performs simulation of the virtual environment. Some proposed distributed virtual environment systems perform such simulation without a server and instead use individual clients. However, the system we are developing performs the simulation with a server in order to maintain the consistency of the environment.

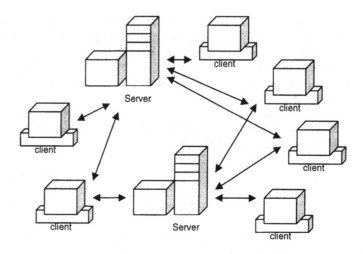

Figure 1: Client server relationship.

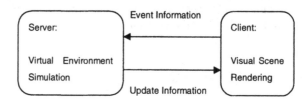

Figure 2: Information exchange.

Clients act as communication front-ends for their users. A client gets a scene graph, and renders a 3D graphics visual scene based on the scene graph. The client also handles events from input devices such as the keyboard, mouse, and joystick, and sends this information to the server.

2.2 Processing Timing

In distributed virtual environments, the service is provided on as a real-time basis. Therefore, both the simulation process in a server and the rendering processes in clients are executed as real-time processing. In order to display a visual scene smoothly, both processes are performed cyclically and continuously more than ten times in a second. This is to say that both processes are performed repeatedly in a period, and the period is shorter than 100ms. We call this period a "Tick".

In a server, a simulation process is cyclically performed at every Tick. In this process, the effect of event information provided by the clients is also considered. After every simulation process, information that has changed in a virtual environment such as position, and orientation of objects, is picked up. This information is distributed to clients as update information.

In a client, the update information provided by the server is used to reflect the state of the server to the client. After the reflection process, the client renders an updated visual scene. These processes also performed cyclically and continuously at every Tick.

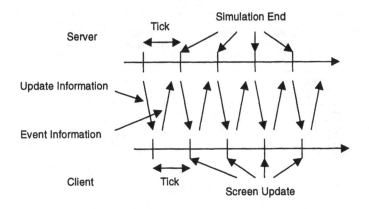

Figure 3: Processing timing.

3 Scene Graph Distribution

In this chapter, we discuss scene graph distribution methods.

3.1 Scene Graph

In this section, we review a scene graph [Hartman96, Wernecke94, Sense8-96], which we focus in on this paper.

This scene graph was originally designed for 3D graphics systems, and is now an essential form of technology for real-time 3D graphics systems, especially for virtual reality systems. The scene graph contains data about all objects in a virtual environment, such as their shape, size, color, and position. Each piece of information is stored as a node in the scene graph. In the scene graph, these nodes are arranged hierarchically in the form of a tree structure. The 3D graphics system traverses the tree structure to extract information and renders the appropriate scene based on the information.

We show an example of a scene graph in **Figure 4**. This scene graph describes a room, and the room might be a subpart of a larger scene graph. The scene graph for the room shows that there are four objects in the room. The node described "Group (Desk)" shows the visual description of the desk. This scene graph shows that the desk is constructed from two parts, one is a desk top and the other is legs. The scene graph also shows that both parts are constructed from their sub parts. The scene graph of the desk is traversed by a rendering system to render a visible scene.

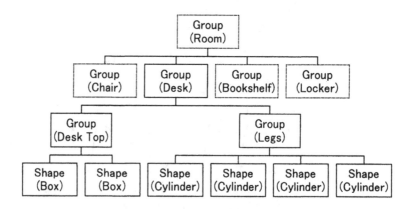

Figure 4: An example of scene graph.

In the scene graph, the Group nodes show what the parts of the objects are, so the nodes have some sub-nodes. The Group node has information that shows:
- Position.
- Bounding box.
- Sub-node list.

The bounding box specifies an outline size of the object. The bounding box is a box that encloses the shapes in a sub-node of the node, and it is used to provide a visibility evaluation of the object to be described later.

The Shape nodes show what the visual shapes of the object are, and the nodes become terminate nodes. The Shape node has information that shows:
- Shape.
- Size.
- Color.

In the VRML[Hartman96], which is widely used scene graph, the Group node corresponds to a Group node and a Transform node, and the Shape node corresponds to a Geometry node and an Appearance node.

3.2 Scene Graph Distribution Problem

There are some options to implement a scene graph distribution method. In this section, we discuss two major distribution methods.

3.2.1 CD-ROM Basis

Scene graph distribution by CD-ROM is appropriate for large virtual environments, because the capacity of a CD-ROM is quite large. Some service providers of distributed virtual environments choose this distribution method for their service. The advantage of this method is that there is no need to distribute a scene graph at each connection time.

However, this distribution method has a drawback in that the person who wants to present a service using a distributed virtual environment must publish the CD-ROMs which contain the scene graph of his or her virtual environment. This restriction makes it impossible for ordinary people to publish their own virtual environment material on the network. Also, a person must publish a new CD-ROM when he or she changes his or her virtual environment. In addition, users may not wish to have to keep many CD-ROMs for individual distributed virtual environment services.

3.2.2 Network Basis

Scene graph distribution by network is well known as the method for VRML. This method has advantages in that anyone can access every VRML scene graph provided on a WWW server, and anyone can get the latest version of the scene graph. On the other hand, the method has disadvantages in that during a scene graph distribution. The user has to wait for the completion of a distribution, and usually cannot see the visual scene of the scene graph until this is done.

For example, the amounts of scene graphs used by some service providers of distributed virtual environments are from 1M bytes to 20M bytes. These amounts indicate the compressed sizes of scene graphs. The download time for 1M bytes scene graph is at least 6 minutes and the download time for 20M bytes is at least 2 hours, in case of a user uses a 28.8K bps modem.

In order to reduce the waiting time, a method called Inline is introduced in VRML scene graph. The Inline method provides an ability to divide a scene graph into a main scene graph and segments. In this method, a user can download the scene graph, segment by segment, and a visual scene is immediately displayed on screen after the main scene graph is downloaded. Therefore, a user's waiting time is reduced if the size of main scene graph is maintained small.

The Inline method, however, has several disadvantages. The order of the downloading of segments for instance is not considered, so the order is usually inefficient for visual effect. The download completion time also becomes lengthened, because each downloading of segments expends individual connection setup time. In addition, extra work is required by an author of a scene graph in order to divide the scene graph into its parts effectively. Moreover, the VRML has a facility to share a sub-scene graph with some scenes. However, the Inline method disables the facility between divided segments, so the total amount of the scene graph usually becomes lager.

3.3 Requirements for Scene Graph Distribution

First of all, we believe that it is very important that everyone be able to publish his or her virtual environment easily. In addition, everyone should always have access to the latest versions of virtual environments. In order to satisfy these requests, the distribution method of scene graphs must be on a network basis instead of on a CD-ROM basis. Second, from the user's point of view, the current waiting time for the downloading of a scene graph at the start of a session must be avoided. Third, from the system's point of view, the amount of scene graph distribution in a session must be reduced in order to reduce network traffic. In addition, processing loads must be distributed to clients as a whole instead of concentrated on a server.

4 Incremental Scene Graph Distribution

In this chapter, we propose the method of an incremental scene graph distribution.

4.1 Overview

In order to satisfy the requirements discussed in section 0, we propose an incremental scene graph distribution method, and this method has these features:
- Incremental segmented scene graph distribution.
- Partial scene graph distribution.
- Visibility importance order distribution.
- Visibility importance that is evaluated by each client.

4.2 Visibility Evaluation

In our method, scene graph segments that express visible objects are distributed incrementally from a server. On the other hand, scene graph segments that express currently invisible objects are not distributed. In order to realize such selective distribution, an evaluation of the importance of an object's visibility is very important.

4.3 Basic Method

The visibility evaluation process is known as culling. This culling process is performed as follows. First, it calculates a viewing frustum (**Figure 5**) which shows users a viewing area in a virtual environment. Next, the process checks that which objects are in the frustum and which are not. The objects in the frustum are visible because they are located in the viewing area of the user. The objects outside of the viewing frustum are, on the other hand, invisible, because these objects are not located in the viewing area of a user.

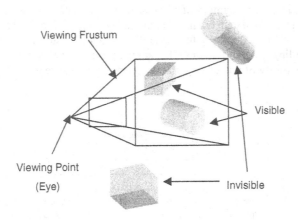

Figure 5: Visibility evaluation.

4.3.1 Server Side Visibility Evaluation

The culling process usually needs information regarding what objects are present and where they are located. This information is described by a scene graph. The whole of the scene graph is managed by a server, so it is very natural that the visibility evaluation is performed by the server. However, this method has a serious drawback.

This visibility evaluation processing requires extra processing costs of the server because the standard processing of the server does not include such evaluation. In addition, visibility evaluation processing is required for each client that is currently connected to the server. This means that the cost of the evaluation process becomes extremely expensive when many clients are connected to the server at a time. This extra processing cost uses almost all the processing power of the server, causing the inability of the server to perform the simulation of the virtual environment.

4.3.2 Client Side Visibility Evaluation

The culling process is a basic process of scene rendering in which each client performs the culling in each rendering process. Therefore, clients can perform a visibility evaluation without any extra processing cost. In our method, we use the information produced by the rendering system to perform a visibility evaluation.

This method has strong advantages in that a processing load can be distributed to each client, and there is no extra load for the server and clients. However, this method has a paradox to perform. The paradox is as follows: The visibility evaluation is performed to choose what scene graph segments should be distributed. It means that the information of the chosen segments is not placed in the client, but the evaluation must be performed by the client using the segment's information.

In order to break the paradox, our method divides a node information into two categories, one has a sub-node list, the other has position, bounding box, and color. The server distributes a segment of a scene graph in a form illustrated in **Figure 6**. The segment is constructed from a parent node and its immediate sub-nodes. A parent node has information about sub-node list, and the sub-nodes have their own position, bounding box, and color. This segment form enables the culling system to evaluate the visibility of a node, and the rendering system to render a temporary shape to a screen.

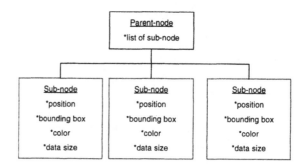

Figure 6: Distribution unit of a scene graph.

All of scene graph segments are distributed according to the requests from a client. However, a visibility evaluation process requires a root node to start the process. Therefore, only the root part of a scene graph is actively distributed by a server exceptionally when a client connects to that server.

4.4 Request Priority

According to a visibility evaluation, there are objects chosen which should be distributed to a client. As well, scene graph segments of these objects are requested by a server. However, all chosen objects might not be distributed at one time, because the network bandwidth is limited. Therefore, we should consider about the distribution order to achieve the distribution effectively. In our method, the objects are distributed according to their visible importance and their distribution efficiency order, as shown below:

- The object that is located near a user in the virtual environment is important
- The object that has a large visible size is important
- The object that is described by a small amount of data is efficient for distribution

The distance of an object can be calculated from the object's position. The visible size of an object can be calculated from the object's distance and bounding box size.

The distribution efficiency of an object can be calculated from the object's visible size and data size of immediate sub-node of the node.

The process flow of clients is shown in **Figure 7**.

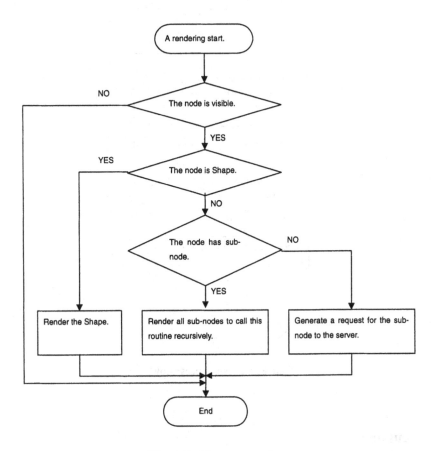

Figure 7: Client process flow.

4.5 Distribution

According to a request from a client, a server will distribute segments of a scene graph. The segments requested by the client are distributed by the server which complies with the order in the request. The distribution for each request is performed during a time that corresponds to a Tick, because new requests will come from the client at every continuous Tick. The process flow is shown in **Figure 8**.

After the distribution, if some segments are remaining in the request, the request for segments is simply discarded. If the segments in discarded requests are important for a continuous visual scene in the client, a request for the segments can be generated by the client again. In a request, the priorities of the segments might be increased if segments that have higher priorities are already distributed according to the previous request; the request for a higher priority of segments would then be removed.

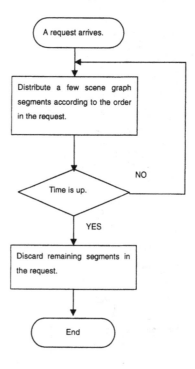

Figure 8: Server process flow.

5 Discussion

In this section, we illustrate a scene graph distribution and discuss the advantages of our proposed method.

Initially, only the root part of a scene graph is actively distributed by a server when a client connects to that server. The root part is constructed from a complete root node and simplified immediate sub nodes of the root node. In this example, the root node is for a room shown in **Figure 9**. A dotted lined box expresses a simplified node. A solidly lined box expresses a complete node. The information stored in each node is the same as that of **Figure 6**.

According to the distributed root part of the scene graph, a rendering system of a client tries to render a visual scene. Here, we assume that only the desk is in a user's viewing area. The rendering system tries to render the desk in detail, but the information already provided by the server is not enough to render it in detail. The rendering system then tries to render the desk based on the information that is provided by the simplified node. This simplefied node has information such as position, bounding box, and color. In this situation, the desk is rendered as shown in **Figure 10**, (A). Thus, a user cannot see the detailed shape of the object, but can understand that something is there. As shown in here, the user can see an outline of an environment immediatery after connection.

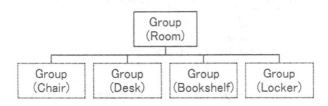

Figure 9: Initially distributed scene graph.

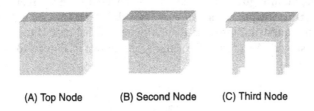

(A) Top Node   (B) Second Node   (C) Third Node

Figure 10: Incremental desk rendering.

The client can then request more detailed information about the scene graph to the server. According to the request, the server distributes the detailed part of the scene graph (**Figure 11**) of the desk to the client. The client grafts the scene graph into an existing scene graph. **Figure 12** shows the result.

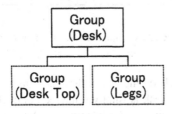

Figure 11: Scene graph segment.

As shown above, our method divides a scene graph into these segments automatically, and distributes the segments incrementally according to requests from clients. In this manner, scene graphs in clients are developed continuously, and the appearances of objects that are seen by users are developed as time goes on.

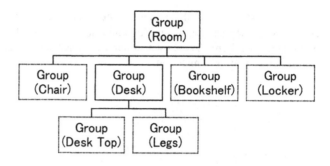

Figure 12: Scene graph grafting.

In the next rendering cycle in the client, the rendering system uses the grafted scene graph shown in **Figure 12**, so the desk is rendered from the scene graph shown in **Figure 10**, (B). Now, the user can guess that the object may be a desk or a table. The client then requests information about a more detailed part of the scene graph of the desk to the server. As a result, the scene graph shown in **Figure 13** is constructed.

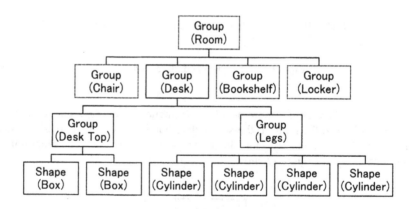

Figure 13: Completed description of a desk.

In the next rendering cycle on the client, the rendering system uses the scene graph shown in **Figure 13**, so the desk is finally rendered completely from the scene graph shown in **Figure 10**, (C). Now, the user cleary understand that the object is a desk.

As shown in **Figure 13**, our method distributes only the required parts of a scene graph. This means that our method never distributes unnessesary parts of the scene graph. This method realizes an efficient network bandwidth utilization. In addition, the distribution of nessesary parts of the scene graph would be accelarated, because the distribution of unnessesary parts of the scene graph are automatically avoided.

6 Conclusion

In this paper, we proposed an incremental scene graph distribution method for distributed virtual environments. The method realizes an efficient scene graph distribution in narrow bandwidth network.

References

[Maxfield95] Maxfield, J., Ferrence, T., Dew, P.: "A Distributed Virtual Environment for Concurrent Enginnering", In Proceedings of VRAIS '95, IEEE, (1995).

[Broll95] Broll, W.: "Interacting in Distributed Collaborative Virtual Environments", In Proceedings of VRAIS '95, IEEE, (1995).

[Stansfield95] Stansfield, S., Shawver, D.: "An Application of Shared Virtual Reality to Situational Training", In Proceedings of VRAIS '95, IEEE, (1995).

[Barrus96] Barrus, J., W., Waters, R., C., Anderson, D., B.: "Locales and Beacons: Efficient and Precise Support For Large Multi-User Virtual Environments", In Proceedings of VRAIS '96, IEEE, (1996).

[Kessler96] Kessler, V., D., Hodges, L., F.: "A Network Communication Protocol for Distributed Virtual Environment Systems", In Proceedings of VRAIS '96, IEEE, (1996).

[Macedonia95] Macedonia, M., R., et al.: "Exploiting Reality with Multicast Groups: A Network Architecture for Large-scale Virtual Environments", In Proceedings of VRAIS '95, IEEE, (1995).

[Singh95] Singh, G., et al.: "BrickNet: Sharing Object Behaviors on the Net", In Proceedings of VRAIS '95, IEEE, (1995).

[Hartman96] Hartman, J., Wernecke, J.: "The VRML 2.0 Handbook", Addison Wesley, (1996).

[Wernecke94] Wernecke, J.: "The Inventor Mentor", Addison Wesley, (1994).

[Sense8-96] Sense8: "WorldToolKit Reference Manual, Release 6", (1996).

Value-Added Services in Industrial Automation

André Hergenhan[2], Christoph Weiler[1], Karlheinz Weiß[1], and Wolfgang Rosenstiel[12]

[1] Forschungszentrum Informatik (FZI) Karlsruhe, Haid-und-Neu-Straße 10-14, D-76131 Karlsruhe, Germany
[2] Universität Tübingen, Technische Informatik, Sand 13, D-72076 Tübingen, Germany

Abstract. Many of todays existing industrial applications are controlled by a so-called embedded system, which is application-specific hardware with a specific software running on it. To monitor, control, maintain and update the system, the presence of an expert is highly recommended. Remote monitoring and control would provide a considerable advantage for such industrial applications. In order to do so, the increasing number of communication possibilities via the Inter- or Intranet would have to be made available in a simple, inexpensive and secure manner. This paper describes a generic approach, where the up-to-date standard techniques *Java* and *Actuator Sensor Interface* (ASI) are combined in order to support a simple integration of arbitrary industrial applications on the Internet.

1 Introduction

1.1 Embedded systems and the *Actuator Sensor Interface* (ASI)

One of the key tasks of embedded systems[1] in the field of industrial automation is to gather information about the current state of a technical process via sensors, evaluate this information in order to control the behaviour of process events and therefore influence the running process using appropriate actuator devices.

The majority of these sensors and actuators are only binary, that means, sensors have only two states like open/closed (e.g., a simple switch or light barrier) and actuators are also limited to two states, like active/not active (e.g., a motor switched by a relay).

In modern plants, systems with sometimes thousands of binary sensors and actuators must be realized. The traditional method of connecting these devices to a control unit is the star wiring topology. That means, every device is connected to an appropriate I/O unit with its dedicated cable. This method results in large and expensive cable trees and a time consuming installation process.

---

[1] This work was supported in part with funds of the "Deutsche Forschungsgemeinschaft" under reference number 3221040 within the priority program "Design and design methodology of embedded systems"

In recent years, the progress in microelectronics enabled a new method called *Actuator Sensor Interface* (ASI), which is becoming more and more popular. The key feature is the possibility to connect up to 128 binary sensor and actuator devices with the control unit using only one simple bifilar cable. This cable is used for both power supply and information exchange.

A single microchip called ASI slave groups up to four sensors and actuators, connects them to the ASI cable for power supply and watches for communication calls. These calls, which are cyclic transmitted from an ASI master, consist of a serial communication protocol. The ASI slave chips receives the output data and forwards them to the actuators, if the address matches.

Additionally, the selected slave chip answers immediately and transmits input data from the sensors back to the ASI master. Up to 32 ASI slave chips can be connected to one ASI cable, which enables a free tree wiring topology. This configuration is called ASI channel (for details see [1] and subsection 4.1). The use of ASI leads to remarkably reduced costs, reduced installation efforts and to a higher reliability of the embedded systems.

1.2   Internet and *Java*

An increasing number of modern communication systems use the fast growing Internet as basic communication platform.

Connecting to the Internet is possible from almost every place in the world and in the same way almost every place can be reached by using different Internet-based connection services.

A common communication platform is the TCP/IP protocol, which is the base of our work, too. Fig. 1 shows the relevant protocol layers of the Internet. The shaded areas indicate the protocol layers which are relevant for the work presented in this paper.

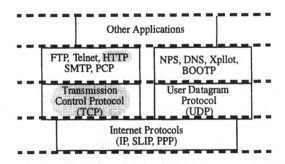

Fig. 1. Internet protocol layers

In connection with the growing Internet, *Java* has become a standard programming language in the domain of Internet programming, due to its platform-independence.

Platform-independence was gained by defining the *Java Virtual Machine* (JVM) [2], a standard code format (bytecode) and a standard format for the executable files (class-files). This standard code format, wrapped in class-files, can be processed on each platform where a JVM realisation exists.

The possibility of having a standard programming language and a standard executable offers several advantages, if a designer wants to integrate embedded applications on the Internet. Now, an application can be monitored, controlled and updated via Internet.

The same programming environment could be used to create a client/server architecture in order to design a remote control system. Further, the connection to an Intranet or to the Internet is supported in a convenient way by standard *Java* packages.

The executable bytecode generated can be processed by each realisation of the JVM. But, the realisations strongly differ in their performance and in their costs. Fig. 2 shows three basically different possibilities to execute the bytecode.

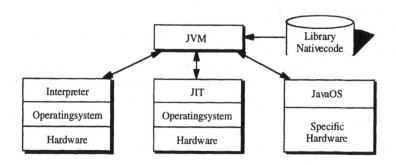

Fig. 2. Alternative realisation possibilities of the JVM

The most common way to execute the bytecode is the simple interpretation of the bytecode sequence.

A second possibility is to translate the bytecode into the processor's native code at the first invocation of the code, which is called a *Just-In-Time* (JIT) compilation. This improves the performance, if code parts are executed frequently, so that the overhead invoked by the compilation process can be ignored.

The most efficient method of bytecode execution is the use of specific hardware, e.g., the *picoJava*[2], which has a bytecode-oriented instruction set and the processing model defined in the JVM.

Many variations exist for each of the three alternative realisation groups. Depending on the system under developement, an appropriate realisation of the JVM must be selected.

---

[2] developed by Sun Microsystems ™

1.3 Added value by combining available technologies

The combination of the existing standard technologies introduced in subsections 1.1 and 1.2 offers a generic platform for the integration of industrial applications into a network. It then supports the desired services like monitoring, control and update of the embedded system.

The improved functionality of the industrial applications is achieved, since many tasks can be done from a remote location. Additionally, experts have a simple, fast and world-wide access to their applications, which remarkably reduces service costs and improves the availability of the embedded systems. In addition, the Internet is supplied with new capabilities, since new services become available due to the approach introduced in this paper.

1.4 Organization of this paper

Section 2 summarizes our previous work in the areas of ASI and *Java* as well as previous work in the area of remote embedded application control, which was mainly done by other groups. To illustrate Internet-based embedded systems, an example is presented in section 3. The technical details of our generic implementation are explained in section 4. Section 5 deals with security problems on Internet-based embedded systems. We draw some conclusions about our work in section 6 and give a brief overview of future steps.

2 Previous work

Over the last years, we analysed the structure of embedded systems in the field of industrial automation in order to develop systematic design methods. Additionally, *Java* was evaluated, which appears to be the most suitable language to program and connect embedded systems to the Internet.

In the area of embedded system design, several real examples from the domain of industrial automation have been used to evaluate our research approaches. One example was a concept for a multi-channel ASI master, which was developed in cooperation with partners from industry [3].

Several research efforts in our group are concerned with the use of *Java* for designing network integrated systems [4], using *Java* for the creation and initial examination of virtual prototypes [5] and also in the domain of embedded systems [6]. Java offers suitable features for all these different tasks.

Rapid progress in the *Personal Computer* (PC) market influences the field of industrial automation. An increasing number of *Industrial PCs* (IPCs) will be used to replace the traditional *Programmable Logic Controller* (PLC). The most important reason is the open and manufacturer-independent standard in contrast to the closed and manufacturer-dependent PLC implementation.

This fact makes IPC products available world-wide, guarantees a high degree of innovation and low costs. In addition, novel tools like *OPEN CONTROL* simplify communication via modular interfaces in open IPC-based automation systems as described in [7].

When using an IPC, the potential for networking via Intra- or Internet is easy achievable. *Java* programs can be run on the IPC and are mostly used for network management, visual presentations and diagnostic services today.

Like a *Network Computer* (NC) in the office world, many companies work towards a *Network Based Controller* (NBC) [8], which should also be able to realize real-time processing in the near future. As further described in [9], *Java* programs can access *OPEN CONTROL* services and therefore enable remote access features via the Internet.

The methods described in [7, 8] have the advantage of platform independence, but the different layers on top of each other require powerful hardware. So, these methods can only be used in large automation projects.

This is a serious drawback for smaller systems, like many other tasks in industrial automation as well as for the rising market of applications in private households, which should be accessed via the Internet and therefore require low-cost-solutions.

Our approach aims to bring *Java*, which implements the Internet access, closer to the application in order to save computing resources.

Some preliminary research work in this area has been done in [10]. This work considers the problem of integrating native classes and methods into a *Java*-based design while using the *Java Native Interface*. These parts, written in *C*, implement the access to the application-specific hardware. Besides the fact, that the approach does not deal with a real world example, it is, in contrast to our approach, more specific and is not suited for the integration of arbitrary industrial applications.

3 Internet-based embedded systems

3.1 Overview

This section shows our approach of easily enhancing existing embedded systems in order to provide added value services.

To demonstrate and validate the usability of the following proposed approach, we integrated an embedded system on the Internet. The model of an industrial shelf serves as an example. Fig. 3 shows an overview of the complete application.

3.2 The industrial shelf application

Functionality of the embedded system The model has ten columns ($X0\ldots$ $X9$) on the x-axis and five rows ($Z0\ldots Z4$) on the z-axis. The y-axis consists of three positions: $Y(0)$ is used as the drive position, $Y(1)$ is used as an end indication during the movement inside the shelf and $Y(-1)$ is used as an end indication for the movement to one of the two I/O stations. The position $(X, Y, Z) = (0, 0, 0)$ is considered to be located in the most left column and in the top row. For orientation, see the picture in Fig. 3.

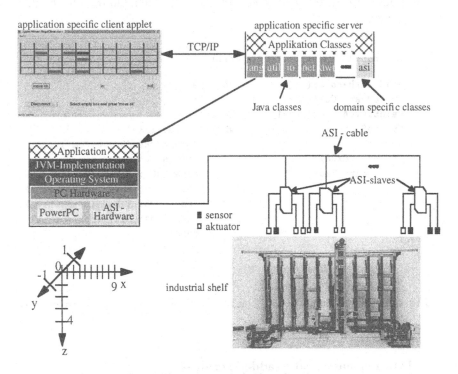

Fig. 3. The industrial shelf application

The industrial shelf is furnished with sensors and actuators which enable the use of ASI for communication. The shelf needs 26 ASI slave inputs for position sensors and eleven actuator outputs for the motors, so seven slave chips are used.

The main task (see Fig. 4) first calls the function *Init()* to initialize all connected ASI slaves and start up communication. The function *Reset()* searches for the current absolute position of the shelf wagon and moves it to the default position $(X, Y, Z) = (0, 0, 4)$.

After that, the main program waits for commands from the Java server running on the PC. If *MoveX(pos)* or *MoveZ(pos)* is issued, the function computes the delta value between the current absolute position and the absolute destination (given in pos). Then it starts and controls the appropriate motor until the shelf wagon reaches the target. From the current position, the command *TakeY()* takes parts out of the shelf and the command *PutY()* puts parts onto the shelf.

The interrupt task is invoked every $150\mu s$ with an new slave answer. The *IntHandler()* first calls the *WriteMasterCall()* function to generate the next master call with the current actuator output data, then it reads back the slave answer from the last master call using *ReadSlaveAnswer(adr-1)* with the current sensor input data. *UpdateSlaveImage(adr-1)* writes the input data to the current process image.

interrupt task   main task

```
void IntHandler() {            void main(){
WriteMasterCall(adr);          Init();
ReadSlaveAnswer(adr-1);        Reset();
UpdateSlaveImage(adr-1);       while (Forever){
adr++;                             command=ReadServerCom();
if (adr==0x08)                     switch (command) {
     adr==0x01;                    case 0x01: MoveX (pos); break;
}                                  case 0x02: MoveZ (pos); break;
                                   case 0x03: TakeY (pos); break;
                                   case 0x04: PutY (pos); break;
                                   default:
                                   }
                               }
                           }
```

Fig. 4. Software structure for the embedded PowerPC controller (written in *C*)

3.3   Internet-based value-added services

The Internet's data-transport protocols provide functionality for data communication by using a client/server architecture [11, 12]. These services are currently implemented in the programming language *Java* on a PC communicating with the embedded system.

The application-specific server The server uses predefined interfaces, which are introduced in sections 4.2 and 4.3, to communicate with the applet and with the embedded system (PowerPC board). The execution model of the server is described in detail in section 4.3.

The server receives instructions from a control applet by establishing a TCP/-IP communication channel. The instructions received are application-specific. The server recognizes the instructions, splits these instructions into simple one, which can then be taken to the embedded system in order to be executed. In the opposite direction, the server receives information about the process image or error states from the embedded system and sends them back to the client.

The server is implemented in *Java*. Due to the fact that *Java* and applets written in *Java* have some security restrictions, it must not be possible to access machine resources, like memory regions which make communication links. To overcome this drawback and to support write and read methods to/from the memory mapped interface needed for communication, native methods implemented in another programming language like *C* have to be invoked within Java.

The *Java Development Kit* defines the *Java Native Interface* naming and calling conventions so that native library methods (dynamic link libraries on Windows-based systems, as well as shared objects on Unix systems) can be invoked.

The application-specific client applet The applet is running in a web browser and mainly consists of an application-specific user interface. In case of the given example, the user gets a representation of the current state of the industrial shelf and an appropriate user interface to send instructions to the application. Here, these instructions are *in, out and move*. Now, a user can select an empty box from the industrial shelf and press the *in*-button, which causes the applet to send the instruction *in(parameter)* to the server. Accordingly, an occupied box from the shelf can be selected and the *out*-button can be pressed to send the instruction *out(parameter)* to the server. A *move* instruction can be sent to the server in a similar fashion. Whenever a transaction is completed, the applet displays the actual state of the industrial shelf and accepts new instructions or allows the disconnection of the applet.

The embedded web server A http-server, running on the PC, provides the interface to get a web page including the application-specific applet.

4 Implementation

4.1 The control of the industrial shelf and ASI

Fig. 5 depicts the different components of an *Actuator-Sensor-Interface* (ASI) system.

The ASI master is connected with up to 32 ASI slave chips via a bifilar ASI cable. The ASI power unit supplies the ASI net with $30V$ DC and $2A$. The ASI master generates a serial master call composed of a five bit address part (A4..A0) and a four bit data part (I3..I0). The single ASI slave chips are distinguished from each other with their own addresses, implemented in a small ROM area on chip and ranging from 0x0 to 0x1F. If the transmitted address part matches the address of an ASI slave chip, the selected slave forwards the four bit input data (I3..I0) to its outputs and transmits immediately the state of the four current input signals back to the master (in the slave answer (I3..I0)).

The time for the transmission of one serial bit is $6\mu s$ and the total time for a master call and slave answer is about $150\mu s$. The master increments the address with each master call and is able to address up to 32 slaves in a sequential manner. When the master reaches address 0x1F, it resets the address to 0x0 and continues with the next polling cycle. In this way, the master is able to generate a new process image every $5ms$. This time is short enough for most real-time applications in the field of industrial automation.

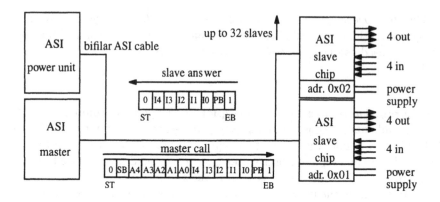

Fig. 5. Schematic diagram of the *Aktuator Sensor Interface* (ASI)

Fig. 6 depicts the hardware architecture of the ASI master. A 32 bit microcontroller (IBM PowerPC 403) acts as the main embedded controller and communicates with a *Peripheral Interface Controller* (PIC 16C57) via a bi-directional pipeline structure, implemented in a *Field Programmable Gate Array* (FPGA XC4010E).

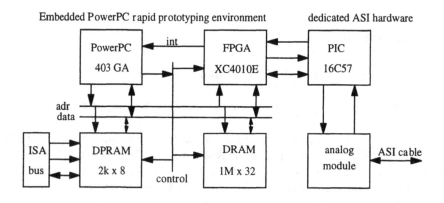

Fig. 6. Hardware architecture of the ASI master

Additionally, the analog module receives the slave answer, which is modulated in the slave chip as well. It demodulates the signal and sends it back as a digital input signal to the PIC. The PIC receives the serial slave answer, decodes the manchester code to a binary code, writes the answer into the pipeline to the PowerPC 403 and signals the appearance of the slave answer with an exception signal (interruption).

The PIC is used for the low level ASI communication task and serves for the hard real-time conditions of the low-level ASI communication task.

The 4 MByte DRAM bank serves as main memory and the 2 KByte Dual Port RAM (DPRAM) as a communication medium with the PC.

This architecture is implemented as a PC add-on board using a rapid prototyping environment [13] in conjuction with a dedicated ASI hardware module.

4.2 The interface between PC and embedded system

Hardware A Dual Port RAM (DPRAM) with the size of 2 KBytes serves as a communication link between the embedded system (the embedded PowerPC environment) and the PC. At PC side this 2 KByte of memory can be arbitrarily moved within the overall memory space from address 0xC0000 to 0xD0000. At PowerPC end the DPRAM is assigned to an appropriate memory bank.

Software Since the application server is running on the PC, it is up to the server to control and to synchronize communication to the lower layered software parts running on the PowerPC. After receiving an instruction from the applet the application server splits these instructions into more elementary ones and sends then opcode and arguments via the DPRAM to the PowerPC environment. The PowerPC platform itself takes this as input doing the called action, providing a new state of the application or even an error code.

Ensuring data integrity, the layout of the DPRAM has been virtually splitted into four parts as shown in Fig. 7.

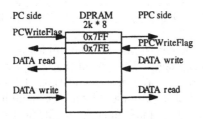

Fig. 7. Memory partitions of the Dual Port RAM

For the sake of simplicity, no interruption driven synchronization of the communication has been introduced, so it has been done by polling handshake flags on both sides.

With this prerequisite, the inner communication protocol has been implemented. The server on the PC can only write opcode and arguments into the appropriate memory fields, if the *PPCWriteFlag* is reset by the PowerPC subsystem. By setting the *PCWriteFlag* the server on the PC signals that a command is pending. The PowerPC should read the instruction within a given amount of time otherwise a timeout error occures. After finishing the requested task, the PowerPC writes the current state into the DPRAM and sets the *PPCWriteFlag*, so it can be read by the PC.

4.3 *Java*-based remote control

The *Java*-based remote control part of the system consists of three parts; the control applet, the server, which offers a connection port, and the communication protocol between the client applet and the server application. All parts are implemented in *Java* and therefore platform-independent. The simple generic structure of the control architecture is shown in Fig. 8.

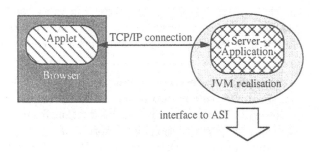

Fig. 8. Generic remote control architecture

Applet The applet always consists of a GUI part, which enables communication with a user. The design of the user interface depends on the industrial application to be controlled.

The base for the implementations are the applet provided *frame-* and the *awt*-classes of the *Java* packages. The second basic feature of the applet is its ability to connect to a server program. To do this, the applet must be aware of the host where the server is running. This is the same host where the applet was uploaded. In addition, the applet must know the port number offered by the server program.

With that information, the applet can create a socket connection to the server. Each socket has an *Input-* and an *OutputStream*. After establishing the connection to the server, a new thread (observer) is generated, which checks with a selectable frequency if new data from the server is available. A flag indicates the arrival of new data and the designer can evaluate the new data according to the current application.

Besides the construction of the connection and the observer, further generic features are the simple user information line of the applet and the ability to disconnect from the server. Based on these general abilities, a designer implements his own application specific control applet.

Server A generic frame for the server architecture shown in Fig. 8 is offered in the same way as for the applet.

In contrast to the applet, the server creates a *ServerSocket* and assigns a port number to this. The applet must be aware of this port number in order to connect to the port. Fig. 9 shows the simple generic kernel of the server.

```
while (everythingOK) {
        createConnection();
        while (connected) {
                receive ();
                execute ();
        }
}
```

Fig. 9. Generic kernel of the server

The method *createConnection()* waits for a client request to connect to the server. This is done by calling the *accept()* method of the created *ServerSocket*. The communication is blocked until a connection is requested. When a client connects to the port, the method *accept()* returns a socket. Then the *Input-* and *OutputStream* of this socket are used for the communication with the applet.

The execution sequence of the server is kept as simple as possible. As long as the connection is correct, the *receive()* method reads in new data from the connection. When new data is available, the *execute()* method is called.

The designer inserts application specific code into the *execute()* method. The instructions received from the client are examined in the *execute()* method and appropriate actions are performed, which includes sending of an answer or several information strings to the applet. The *execute()* method provides at steady connection, so that the next input from the client can be read in.

Protocol Strongly related to the introduced client and server architectures is the protocol used for communication between them.

Two main demands must be fulfilled by the protocol. It must be simple and it must be flexible to satisfy all requirements. The protocol is based on TCP/IP and uses, as already mentioned, the sockets provided in the *Java* net package. These sockets have *Input-* and *OutputStreams*, which offer *read()* and *write()* methods.

The manner in which a connection is established was already explained in conjunction with the client and the server. Since only a limited number of data is transferred between the client and the server and in order to be general, strings are used for the data exchange. The first message sent from the client to the server is a start instruction. The server responds with an *ok* and an arbitrary, application-specific state description. If this communication works, the client can send application-specific instructions to the server.

The server responds using the keywords *ok, state, error,* which is a reasonable abstract frame for communication. Additional keywords for the termination of the communication connection are *stop* and *stopOk.* Each of the strings sent from one point to the other must be leaded by one of the keywords, otherwise a warning is produced by the server or the applet. The basic communication structure and all relevant header keywords are depicted in Fig. 10.

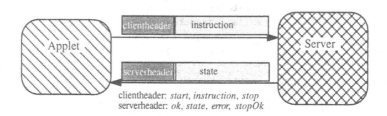

clientheader: *start, instruction, stop*
serverheader: *ok, state, error, stopOk*

Fig. 10. Basic communication structure

5 Security of Internet-based embedded systems

5.1 Overview

Despite all their opportunities, adding internet services to an embedded system also submit a variety of potential security threats. Perfect security is not achievable. There is a strong need of knowledge of weak spots and tools to protect a system from intentional attacks.

Attacks could be made on

- the transfer channels and associated data and
- the executable code running on the embedded system [14].

Security sanctions, like a restrictive permission mode administration, restrictive access control or firewalls provide preventive protection. However, measures like backups, encryption of sensible data, auditing of intruders and assurances in case of events diminish the consequences of security violations.

5.2 *Java*-based security

The *Java* platform was built as a network-save computing model [15]. Its security features are an equally essential part of the package. *Java* provides an integrated security concept with different instances of security.

Java runs in a *Virtual Machine* (VM), which distinguishes local code and remote downloaded code with different permission modes on system resources. Applets e.g., run in a so-called sandbox with restricted access to resources.

The programming language *Java* itself does not support pointer arithmetic, so that arbitrary memory access is impossible. In addition, the JVM provides a bytecode verifier which refuses the execution of illegal program code.

Due to the fact, that we do rely on our own source code, code integrity is given. However, the security of the transfer channels and their associated data is important to the considered class of applications. This leads to satisfy the three main objective of data security: confidence, integrity and availability.

Although it is possible to provide mechanisms of dynamic socket arrangements, it can not be guaranteed, that sensible data are not wiretaped. Therefore, data encryption is of greatest importance. The *Java Cryptography Extension* software package is a standard extension to the *Java Development Kit* software (JDK). It is a comprehensive crypthography toolkit that creates a standard set of application programming interfaces (APIs) for advanced *Java* software-based encryption technologies [16].

There are some problems to ensure data availability over a TCP/IP connection. Due to this fact, time-critical communication cannot be established. However, the application can provide handshaking mechanisms, in order to determine, whether a response of a calling command has occured or not.

5.3 Security on embedded web server

The embedded web server is an additional feature of an embedded system [11, 12]. Security within this embedded web server means that the access to sensible information respective configuration data is restricted. Instead of many people only a few have access (by password) and permission (by setting permission modes) to get, put or change data and code to or from the web server running on the embedded system.

6 Conclusions and Future Work

6.1 Conclusions

This paper introduced a generic approach to the integration of industrial applications on the Internet or an Intranet. Integration in this context means remote monitoring, control, maintainance and updating of the application.

To make these features available in a convenient way, the approach combines two established up-to-date technologies, *Java* and ASI, to provide a comprehensive but also generic frame.

The software part of the generic frame currently consists of the *Java* part, an applet and a server application and a *C*-based interface module to access special functionality running on a PowerPC hardware platform.

The hardware part of the approach currently uses a PC to run web server and the server application. The PC is equipped with a PowerPC add-on board which is connected to additional ASI specific hardware.

To prove the usability of our generic approach, an example of an industrial application was integrated into the Internet.

We have shown, that the use of the predefined frame allows for fast and convenient integration of such an application on the Internet.

6.2 Future Work

We will extend our embedded platform with physical links (Ethernet, ISDN) in order to connect the embedded system to arbitrary networks in a simple fashion. This allows a connection between an embedded system and the Internet all over the world.

Further, we evaluate different real-time operating systems with respect to their ability to support a JVM additionally running on the same embedded system. This is a further step towards a compact, powerful and inexpensive embedded (core) system for industrial automation able to communicate via Intranets or the Internet.

References

1. W. Kriesel and O. Madelung. *Das Aktuator-Sensor-Interface*. Carl Hanser Verlag, München, 1994.
2. T. Lindholm and F. Yellin. *The Java Virtual Machine Specification*. Addison-Wesley, Reading, Massachusetts, 1996.
3. V. Douradinova and K. Weiß. Architektur mehrkanaliger Hochleistungs-Master für ein Aktuator-Sensor-Interface (ASI). In *Tagungsband iNet-Kongreß, Karlsruhe*, 1996.
4. C. Trautwein and W. Rosenstiel. Elektronik-CAD-Anwendung im WWW. *Proceedings of CAD '98, Darmstadt*, 1998. (to appear).
5. C. Weiler, A. Kunzmann, and W. Rosenstiel. Performance Analysis for a Java-based Virtual Prototype. In *Proceedings of the Eighth International Workshop on Rapid System Prototyping, Chapel Hill*, 1997.
6. W. Rosenstiel and C. Weiler. Java for Embedded Systems Design. In *Proceedings of SASIMI '97, Japan*, 1997.
7. A. Baginski. OPEN CONTROL - Der Standard für PC-basierende Automatisierungstechnik. *Feldbustechnik in Forschung, Entwicklung und Anwendung*, October 1997.
8. P. Fröhlich and T. Seidel. OPEN CONTROL: Der Weg zum Network Computing. *Feldbustechnik in Forschung, Entwicklung und Anwendung*, October 1997.
9. J. Gosling, B. Joy, and G. Steele. *The Java Language Specification*. Addison-Wesley, Reading, Massachusetts, 1996.
10. H. Reiter and C. Kral. Java und LonWorks - von der Fernsteuerung zur Visualisierung. *Feldbustechnik in Forschung, Entwicklung und Anwendung*, October 1997.
11. S. Wingard. Devices Web-basiert konfigurieren. *Systeme*, 11(11):31–34, November 1997.
12. R. A. Quinnell. Web servers in embedded systems enhance user interaction. *EDN Europe*, pages 61–68, April 1997.
13. A. Hergenhan and K. Weiß. *User's Manual: Microcontroller Design Environment*. FZI Karlsruhe, 1997.
14. N. Luckhardt. Surfer's Hai Security. *c't magazin für computertechnik*, 1997(13):166–174, November 1997.
15. R. Pons. Under Lock and Key: Java Security for Networked Enterprise. *http://java.sun.com*, January 1998.
16. Sun Microsystems. Sun Premieres Java Cryptography Toolkit. *http://java.sun.com*, January 1998. Press Release.

Requirements and a Proposal for the Prevention of a Class of Service Interactions in Intelligent Networks

Dirk O. Keck

Institute of Communication Networks and Computer Engineering (IND)
Prof. Dr.-Ing. Dr. h.c. Paul J. Kühn, University of Stuttgart,
Pfaffenwaldring 47, D-70569 Stuttgart, Germany
keck@ind.uni-stuttgart.de

Abstract. Service interactions in Intelligent Networks are a major obstacle to the fast and cost-efficient introduction of new telecommunication services. In this contribution, a class of service interactions are addressed that result from deficiencies of the IN infrastructure to provide to a service the necessary information to carry out its task properly. A number of requirements are established that remove or at least reduce these deficiencies, and a proposal is made how these requirements could be realized. In this proposal, the concept of *Service Admission Control (SAC)* is introduced, which is based on a session concept represented by a dynamic *Session Configuration Profile (SessCoP)* accessible and modifiable by IN services that may influence a call relationship (session). This SessCoP allows the IN services to gather the information needed to carry out their tasks properly, that are not provided by the messages of the signalling protocols. The application of these concepts are explained by two brief case studies. This contribution reports on work in progress.

1 Introduction

1.1 The Intelligent Network

The Intelligent Network (IN) provides architectural support for the realization of new services in communication networks. It is based on the principle of separation between the basic functionality of the switches and the functionality of the services. This leads to a simplification and acceleration of service development, since it is not necessary to modify the software within the switches to introduce new services, and to a reduction of management effort, as the services are concentrated into service control points. For a more detailed introduction into the concept of the Intelligent Network, the reader is referred to [1, 2].

1.2 The Service and Feature Interaction Problem

The terms of service and feature interaction refer to situations where two or more communications services or service features (as the basic elements of a communications service) influence each other mutually. This results in a behaviour of at least one of the services or features, that is different from the behaviour of this service or feature without presence of the other service or feature. The term *interaction* does not imply that this behaviour is undesired; if this fact has to be expressed, the term *interference* should be used. [3, 4] provide a concise introduction into this problem area.

1.3 Classification of Service Interactions

The cause of a service interaction is an important classification criterion [5]. In this paper, a slightly different taxonomy is used, based on a scheme sketched in [6]. This taxonomy is based on the observation that feature interactions emerge at different levels of abstraction in the view of a service, and that this fact presents a good taxonomy for classification of interactions with respect to dealing with them.

- *Logical Level.* The logical level is used to model concepts like a service's purpose and intention, focused on the service users' perspective. This level corresponds to the service plane in the IN conceptual model (INCM, [7]). At this level, the way how a service is provided, i. e. the assignment of the different functions to different functional entities, as well as the algorithm and the information used in order to provide the services are neglected.
- *Level of the Abstract Architecture.* On the level of the abstract architecture, a service is modelled in a network- or architecture-dependent way, i. e., the mapping of functionalities to functional entities, the algorithms used, and the information needed in order to provide a service are an issue. This level corresponds to the Global and Distributed Functional Planes in the INCM. The main ideas presented in this paper relate to this level of abstraction, more exactly to the way how the appropriate information about the state of the network and suitable mechanisms to get this information can be provided to a service.
- *Level of the Concrete Architecture.* The concrete architecture takes into account all the irregularities of real systems, like, e. g., physical limitations (memory, bandwidth), interworking with legacy systems and services, different signalling protocols and procedures. It corresponds to the Physical Plane in the INCM. Except for the aspect of interworking with legacy services, this level is not within the scope of this paper.

2 Interactions Caused by Insufficient Information

A significant number of service interactions are caused by insufficient information on the state of the network and of other services being involved in a call relationship. This may result in incorrect decisions of the services and in conflicts concerning the access to system resources. Interactions that belong into this class are addressed by this contribution.

2.1 IN Services in Present Systems

The design of IN services for today's networks is mainly based on the following design guidelines, as far as a treatment of service and feature interactions is concerned: (1) A proper interworking with network based services has to be guaranteed; in cases where this cannot be achieved, the activation of the IN service in combination with these network-based services is prohibited in order to avoid problems. (2) Interworking with the known set of other IN services present in the network has to be ensured by designing the services appropriately or, in difficult cases, prohibiting their activation if conflicting services are present. These quite simple guidelines reflect the limited abilities of the IN infrastructure to deal with interactions.

IN services of present implementations are based on the mechanisms provided by the IN capability set CS-1 (or CS-2)[1] and the digital exchanges using the ISUP (ISDN User Part) protocol. The mechanisms to obtain information on the network state and on other services (listed in Table 1) are in most cases sufficient to implement services that interwork with a number of traditional switch-based and ISDN supplementary services.

These mechanisms provide valuable support for dealing with interactions, but there are some severe restrictions, making it difficult to develop a general method for interaction resolution or prevention on top of these mechanisms:

- For an IN service triggered during a call, it is impossible to obtain information if there is another IN service active on the same call relationship, if the latter service has been triggered later. Furthermore, the information which IN service has been triggered previously can be obtained only indirectly by evaluating the *serviceInteractionIndicator* information element of the INAP. In cases where more than one service have been triggered previously, this may be difficult if not impossible.
- The information that may be exchanged between IN services triggered by different SSP is limited to the capabilities of the ISUP indicators and other information elements.
- The actions that can be carried out for dealing with interactions are limited to disabling the common activation of two services – as far as one service has the possibility to recognise that the other one is active, or to influence a service's behaviour based on the information obtained by this service through ISUP/INAP information elements.

The items presented above demonstrate how limited the capabilities of the IN with respect to dealing with interactions are. During the remainder of this paper, the problems are analysed and a number of requirements for capabilities that would improve the situation will be established. Afterwards, a proposal is made how these requirements could be realized in a real network by an extension to the IN concept called *Service Admission Control (SAC)*.

---

[1] The mechanisms of CS-2 with respect to obtain information on the network state are in principle the same.

Table 1. Mechanisms to deal with interactions provided by CS-1 (or CS-2)

MECHANISM	DESCRIPTION
Mechanisms of IN CS-1 to obtain information on the network state	
Detection Points	Provide information to a service about the point during call processing that has been reached.
INAP Information Elements (Q.1218)	*Information on Directory numbers and parties:* dialledDigits, calledPartyNumber, callingPartyNumber (+ additional information), originalCalledPartyID, additionalCallingPartyNumbers
	Information on call forwarding treatment: forwardCallIndicators, redirectingPartyID, redirectionInformation
	Possibility to propagate ISUP indicators to IN services: serviceInteractionIndicator
Control/Monitor relationship	A concept to guarantee that only one service may actively influence a basic call state model at any given time; the number of services that are notified is not limited.
StatusReportRequest/ StatusReport	Request the SSF to report the busy/idle status of a physical termination resource immediately/on change/on a specific status (this message is not 0 in ETSI Core INAP).
Mechanisms of the ISUP to transport information on services between SSPs	
ISUP Service Interaction Indicators	Fixed set of indicators (in forward and backward direction) that allow to influence the reaction on or the treatment of specific events or switch-based services.
Information Elements	Practically identical to the IEs of INAP concerning directory numbers, parties, and call forwarding treatment.

2.2 Analysis of the Problems and Examples

The design of the messages and information elements of ISUP and INAP are based on traditional two-party calls, enhanced with relatively simple call forwarding and number translation services. The concepts that are introduced by new and complex IN services extend the paradigm of traditional services by a number of aspects described below. In order to allow a proper and correct reaction of the service, a view onto these extensions must be obtainable by the services.

Directory numbers and their mapping to physical lines. The traditional *one-to-one* relationship between directory numbers and physical lines is enhanced significantly by IN services. In many services, one directory number is used to reach multiple destinations, possibly in a time- and origination-dependent way. This introduces a *one-to-many* relationship between directory numbers and physical terminations. Conversely, there is a *many-to-one* relationship introduced by services like MSN (Multiple Subscriber Number). *Many-to-many* relationships exist for services like call centers that work for several companies. A number of services need this information in order to be able to react appropriately. Furthermore, there can be a theoretically unlimited[2] number of participants (directory numbers) involved in a call relationship, like in the example described below. The limitation of the information elements conveyed by the signalling messages may lead to an information loss in such cases.

Multiple Call Forwarding serves as an example for this problem. In this example, the information transfer in a chain of call forwarding steps is studied. A calls B, who forwards to C. C forwards to D, etc. If this scenario is closely examined, the result is that the information on the redirection by party C is lost in the IAM received by E, i. e. party E cannot reconstruct the sequence of forwarding steps that has led to this call. It is obvious, that each service that requires complete information on a call's history cannot work properly. The information obtained by A on the history of this call with the CPG (Call ProGress) / ACM (Address Complete Message) ISUP signals is even more limited.

Resources. Usually, it is assumed by IN services that a resource is available for use by a service when it is available in the network. For example, a specific key on the terminal's keypad or the hook-flash signal is used without knowing whether there is another service that reacts on the same stimulus. As a second example, an announcement requiring user action can be mentioned, where it is assumed that a human user is on the phone who is able to perform the required action. If this is not the case (voice mail, "do not disturb" services with PIN entry), a problem is predetermined.

Voice Mail (VM) and Credit Card Calling (CCC), an example presented in [5], demonstrate a conflict in the use of the #-key, which is used by the CCC service for ending the actual call and placing a new one, and by the VM service to trigger a specific action. Clearly, if both services react on the key, this will lead to a severe and confusing interference. In section 4.2, this examples serves as a case study.

Enhanced Concepts. Traditionally, a call relationship is terminated, if all participating parties are on hook again. Some services enhance this concept by introducing a relationship between subsequent calls, like, e. g., recall of the last caller, where it may lead to severe interactions if the second call (the recall) is treated as independent from the previous one, as it is explained in the following example.

Calling Line Identification Restriction (CLIR), Automatic Recall and Itemized Billing, an example by *B. Cohen,* are services that exhibit an interference although individual pairs of these services are free of interactions. Party A, with CLIR active, has called B, with Automatic Recall. B then recalls A, and can find A's directory number on its itemized phone bill, although A has subscribed to

---

[2] In real implementations, there are, of course, limitations that are introduced by mechanisms like counters (e. g., call forwarding counters), or by physical limitations, but these limitations are caused by design decisions, not by logical restrictions.

CLIR which suppresses the presentation of A's identity (directory number). This interaction is caused by the fact, that the recall is treated as an independent, new call, and not as a consequence of the previous call, where CLIR has been active. It could be resolved (prevented) by providing additional information on the attributes of the last caller's number to the Automatic Recall services (in this case, that the number must not be presented). Of course, this attribute must then be recognized by the Itemized Billing feature.

Charging. The traditional charging concept is based on two major principles: (1) The caller is charged and (2) the tariff can be determined by the directory number that has been dialled. This concept has been extended relatively early by concepts like reverse charging (toll-free numbers), split-rate and premium-rate services, which are, however, in principle built using the same mechanisms.

Advanced IN services introduce the need for more flexible models, like a generalized split charging model (where all participants pay for a call), a third party payment possibility, where a party pays for a call in that it is not directly involved, and the possibility for one subscriber to pay for several IN services he uses during the course of the same call, and to be informed online what amount he is charged for these services.

The above list of abstract concepts extended by IN services does not claim to be complete, but gives an idea where a number of problems that are listed under the generic term service interaction, have their origins.

3 Requirements for the Prevention of these Interactions

3.1 Principles of Service Creation for IN

The IN philosophy follows the guideline of independent and parallel service development (service creation) and also concepts of service privacy in a competitive multi-provider environment. A conventional treatment of service interactions, i. e., taking the implementation of the new service and checking its behaviour in the presence of all other services would clearly offend this philosophy, let alone that this would end up in a complexity problem. Therefore, an important general requirement for prevention and resolution methods is that they must support this independent development of new services.

3.2 Functional Requirements

In the previous chapter, a number of weaknesses of the IN concept concerning the possibility to exchange information between individual services are identified as obstacles to react properly on the threat of service interactions. In this section, requirements for the elimination of these flaws are introduced using a stepwise approach with three levels of increasing functionality and complexity. On the first level, two options are described to notify a service of other services active on the same call relationship.

Level 1a: When an IN service is initially triggered, it has to be informed which other services are already active on this call. This requirement is fulfilled to a significant extent in today's systems by the mechanisms presented in 2.1. Provided this knowledge is available, the IN services have to be designed for proper interworking with known other services (IN and switch-based).

Level 1b: It is possible to obtain information on other services active on this call not only when a service is triggered initially, but also during service processing. This requirement is not fulfilled by today's signalling protocols.

In the above two requirements, it is up to the services to react appropriately on each individual other service, of course only within the limitations of network support. This contradicts to the requirement of independent service development, but in a single-provider environment, the importance of this fact is (still) limited. The following option provides a solution to this problem, the concept of a service-independent Service Template that contains static information on a service, a so-called Service Profile.

Level 2: There exists a Service Profile based on a general template where each service can obtain generalized information on the "internal structure" of other services and react properly on the different alternatives found in the template. The contents of these templates are generated during service creation. It is assumed that each service knows what other services are active (like in levels 1a and b), but they do not need special knowledge of the services, as they can access this knowledge in the form of the static Service Profile.

This template based approach is the first step towards a generalized Service Admission Control (SAC) concept that allows to decide whether a service may be executed or not, or executed in a modified or limited manner, based on the service's requirements expressed in the Service Profile. The problem of such an approach based on a static Service Profile is the fact, that information about what the service could do in principle can be obtained, but not about what the service has done actually. This is an obstacle especially for complex services with many alternative execution paths. This limitation directly leads to the following level 3.

Level 3: The concept of a static Service Profile is replaced by a dynamic Session Configuration Profile (SessCoP), where the actions of a service on a call (session) are recorded and the actual state of this session can be retrieved. It is obvious that this session profile has to be accessible by all services participating in the session.

The requirement outlined on level 3 introduces the Service Admission Control (SAC) based on the Session Configuration Profile (SessCoP), a powerful means to react properly on the actions carried out by other services involved in the same call relationship. During the rest of this contribution, it is described how this requirement can be realized within the existing IN infrastructure and exploited in order to prevent (or resolve) service interactions.

The requirements of level 3 in the previous section is built on top of the concept of a session. Here, an informal definition of this concept is given.

> **Definition (Session):** A session consists of a set of parties that take part actively or passively (e. g., as forwarding parties) in a call relationship. These parties must be related to each other by the existence of (active or suspended) connections or their establishment (successful and unsuccessful) and teardown processes. Thus, the set of participating parties may change over time. A party is involved at each point in time in at most one session; this implies the necessity to be able to join two or more sessions into one. A session ends, if all participating parties are idle and no knowledge (memory) of this session exists in the network that is able to influence the future. Service interactions, as far as they are treated in this paper, are always taking place within one session[3].

3.3 Non-Functional Requirements

The functional requirements presented above are complemented by several other non-functional ones, a selection of which is listed below:

- In order to guarantee acceptance of the method, it has to be integrated seamlessly into the established SIB-based service creation methodology in a way that is as transparent to the service developer as possible.
- The impact on the performance of the services (in terms of answer time) as well as the impact on the computational complexity of the services has to be kept reasonably low.
- As far as the robustness is concerned, there must be useful default behaviours for cases where the session information is obviously inconsistent or non-existent.

3.4 Other Issues

Some other issues that have to be taken into consideration are interconnection of networks belonging to different operators (a major issue for a de-regulated telecommunication market), and the inclusion of services provided by alternative mechanisms like service nodes, and customer premises equipment. These topics exceed the scope of this paper.

4 A Proposal for a Methodology

In this section, it is examined whether and how the concepts and requirements for a Service Admission Control mechanism based on a dynamic Session Configuration Profile that have been introduced in the previous section can be transferred into a real IN infrastructure.

---

[3] The author has no knowledge of interaction examples that occur across session boundaries if a session is defined in the above sense.

4.1   Necessary Basic Mechanisms

Notification of other services on initial trigger.   For the proposal made on levels 1a and b, the notification of other services (triggered after the first service) can be achieved using a unique Service Identifier (Service Key), that is conveyed as a new information element by the ISUP messages. During the call establishment phase, there is always an exchange of ISUP messages between all participating parties, making it easy to transport the new information element without introducing additional messages. The Service Identifier (or, if more services are active, the list of identifiers) are sent to the services as an additional information element of the Initial Detection Point (IDP) INAP message.

More difficulties result from services that are triggered during the stable phase of a call (by so-called mid-call triggers), or during the release phase. In these situations, a message has to be sent that would not be needed otherwise, possibly a Call Progress Message (CPG)[4].

Notification of other services during service processing.   The method described on level 1b requires a means to notify a service during service processing in addition to the initial notification of its activation. Service processing in the IN usually takes place in a stimulus-response manner with short response times. The stimuli are the detection point messages emitted when the DPs are reached during call processing. Under the assumption of short response times, it is sufficient to provide the Service Identifiers of new services active on the actual call together with the event detection point messages sent to a service.

Static Service Profile.   In an IN infrastructure defined according to IN CS-1, it is possible to implement this Service Profile described on level 2 as a static data structure stored in a Service Data Function (SDF) database. There, it can be accessed easily by the services. Performance and database load implied by such a solution is an issue that has to be investigated in further research.

Dynamic Session Configuration Profile (SessCoP).   This concept, presented on level 3, is the most complex one and makes up the key issue in the Service Admission Control concept presented in this paper. It is far more difficult to implement than the concepts of levels 1a,b and 2 due to the dynamic evolving nature of the SessCoP. It requires the possibility to uniquely identify a session, and means to locate, access and modify the session configuration profile. The management of the session configuration profiles is also an issue. Proposals for a realization of these mechanisms are presented in the following paragraphs.

The *Session Identifier (Session ID)* is the means to access the SessCoP of a session. Since it is not necessary for the first IN service triggered in a session to take into account interaction with other services (except for conventional switch-based or ISDN services), a network wide (or worldwide) unique Session Identifier

---

[4] The CPG is used, e. g., in conference calling services to notify participants of actions carried out by other participants, like leaving a conference, isolating a party, etc.

is created and assigned when the first IN service is triggered and is propagated to the other parties involved in this session using the mechanisms described in a previous paragraph. The Session Identifier contains information that is sufficient to locate the SessCoP.

The *Session Configuration Profile (SessCoP)* is represented by a data structure containing generic information that is able to reflect the manipulations an IN service is able to carry out. Table 2 displays a list of information needed for the profile, without claiming that this information is sufficient for the intended task.

Table 2. Information of the dynamic SessCoP

Participants	O-/T-BCSMs with physical line IDs as identification
	Dialled Number history (the way to obtain the term. physical line ID)
	Attributes (terminating, non-terminating)
	Number Attributes (time-/origination dependent, CUG, IN number)
Resources	Keys, Hook-flash signal, Voice Channel (in/out/both), ...
Topology	Connections, Connection Attributes (active, suspended)
Charging	Information on charging (for further study)

SessCoP Manipulation: In order to maintain the information stored in a SessCoP, it is necessary to provide a complete set of operations to manipulate this information. This includes operations to manipulate the individual data members, to clear a SessCoP, to join two profiles and to divide two profiles. The operations needed depend of the contents of the profile and are still under research.

Management of the SessCoPs: For each session a SessCoP is created and maintained. As this profile is accessible only by IN-services, i. e. SCPs, it is an issue how to remove profiles whose sessions have ceased to exist. This information is not necessarily propagated to the SCPs, because synchronization only takes place at armed detection points. However, this is an important issue with respect to security concerns, like the protection of personal data.

There are a number of alternatives to keep this information up do date. First, it is possible to design the IN services appropriately, so that all IN services notify the SCP on release. This possibility creates a large amount of signalling traffic (INAP), which makes this possibility quite expensive in terms of signalling, but also guarantees that the SessCoPs are removed correctly. The second alternative is an aging mechanism, which may lead to inconsistencies because there is no limit to the maximal duration of a session. If this aging mechanism is combined with a check of the parties that belong to a session, possibly by a kind of periodic "keep-alive message", this problem could be circumvented. The third possibility is to monitor for all physical line IDs to which session they belong, and if there is no physical line ID assigned to a specific session, it can be removed. This would lead to a quite complicated and error prone algorithm.

Impact on Service Creation: The maintenance of a SessCoP adds a significant amount of complexity to IN service logics. It cannot be avoided that also already existing services will be affected and have to be adapted. However, a large part of this additional complexity can be hidden from the service designer by an appropriate design of the SIB library.

4.2 Case Studies

The case studies described in this section are coarse grained and simplified with respect to the full amount of signalling messages, information elements and internal processing required. The intention is to illustrate the method described in this paper, not to give a complete and unambiguous specification of it. The contents of the SessCoPs printed in *italics* are displayed as far as they are relevant for the actual case study.

Voice Mail (VM) and Credit Card Calling (CCC). In this case study, it is shown how the interaction in the example explained in 2.2 can be resolved by SAC mechanisms using the dynamic SessCoP presented in this contribution. Both services are realized using the Specialized Resource Function (SRF). In Fig. 1, the message exchange during service processing is sketched together with the contents of the SessCoP. Party A calls the CCC number. A session identifier (ID) is assigned. The credit card number and the voice mail service number are entered by the user and the appropriate event notifications are requested by the CCC service, the Session ID is returned. The SessCoP is updated by a *UpdateSessCoP* message. Afterwards, the VM service number is connected and the service is triggered, this time with a session ID. The VM service tries to allocate the necessary resources. The conflict concerning the use of the #-key is detected and reported by a *SessCoPReply* message to the service logic. In this situation, different reactions are imaginable:

1. The VM service is ended with an announcement *"Failed to allocate the #-key resource"*,
2. The VM services request another unreserved key, for example the ⋆-key, which has to be announced to the user, of course.
3. The user of the two services may actively select (by using his display or an announcement) to which of the services the #-key signal will be propagated.
4. There is a concept of a "context" of a service, and a signal is propagated to the service whose context is active, similar to the service architecture described in [8].

In the example shown in Fig. 1, the second alternative has been chosen. The VM service can be quit either by going on hook or by sending a #-key signal to the CCC service. If the latter case takes place, the reserved resources are released. If the CCC call is ended by going on hook, the session is finished. Now the cleanup mechanisms described in 4.1 come into action.

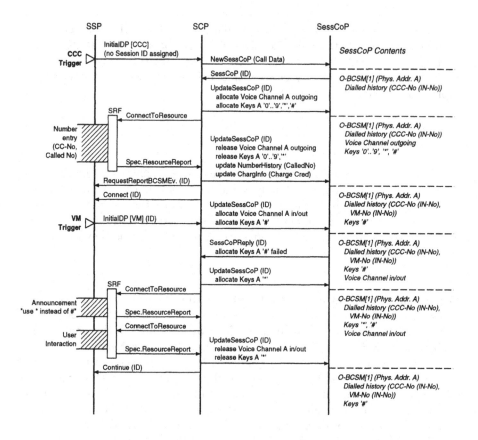

Fig. 1. Interaction of *Credit Card Calling (CCC)* and *Voice Mail (VM)* services.

Dial-In conference with Advanced Originating Call Screening. In this example, it is explained, how a service called Advanced OCS[5] does behave together with a conference call service realized as an IN service on a specialized resource (conference bridge). The conference call service is implemented in a way that a new party that calls any of the parties already involved in the conference is automatically forwarded to the conference bridge. Fig. 2 displays a part of the interaction of AOCS and Conference Calling. For clearness sake, the SSPs of D and A and the SCPs of the both services are drawn separately. Party D (with B on its screening list) calls A, AOCS is triggered and the session ID-AOCS is assigned. When the Conference-Forwarding service of A is triggered, it is detected that there is already a session present (ID-Conf). The both session are joined, and ID-AOCS and ID-Conf reference this joined session. However, one of the IDs (ID-AOCS) is elected for future use, the other one is no longer propagated. After the Answer message (ANM) has been returned to D's SSP, the SessCoP is

---

[5] An Advanced OCS is an originating call screening service that does a second verification before the alerting state is reached or the call is connected (both detection points are available only in IN CS-2).

checked again using D's screening list, recognizing that B is taking part in the Conference. Therefore, D's AOCS service logic releases immediately.

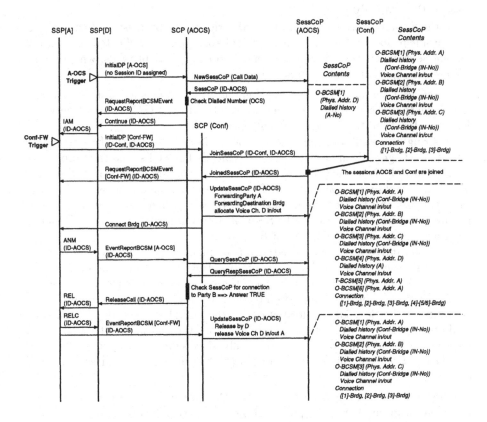

Fig. 2. Interaction of *Advanced Originating Call Screening (AOCS)* and *Conference Calling* services.

4.3 Technical and Realization Aspects

Session ID Assignment. For the assignment of the Session IDs there are basically two alternatives, an assignment in the SSP or an assignment in the SCP.

It is probably more convenient to assign a Session ID in the SCP after the first service has been triggered, and the Session ID is returned together with the first SCP-SSP message sent after the trigger message. This strategy would assign to the SCP the responsibility to maintain the SessCoP, and the Session ID would contain an address of this SCP, together with an ID to identify the profile on this SCP. This strategy is used in the above case studies. However, it leads to problems if the initial trigger is a notification, i. e. has no reply (used in services like televoting). This serious restriction can be circumvented if the assignment of a Session ID is requested explicitly in cases where it has not been already assigned. In these cases, the SSP has to request the assignment of a Session ID using a request as initial trigger, while otherwise a notification is sufficient if a Session ID is already present.

If the Session ID is assigned in the SSP prior to the processing of the trigger, this problem does not occur, but it is probably more complicated to create a unique Session IDs here, as there are much more SSPs than SCPs, which makes the creation more difficult to maintain.

SessCoP Location. The Session Configuration Profiles are data structures, that have to be accessible by the Service Control Points (SCPs) of the network. For their location, there exist different options.

A straightforward solution is a location in the SCP where the first service in a session is triggered. In an IN, there are typically few SCPs, and the probability is high that all IN services that influence one session are provided by the same SCP. This would reduce the inter-SCP information exchange necessary to access the SessCoPs and to keep them consistent. In cases where SessCoPs are joined, it may be necessary to transfer the SessCoP information to another SCP and to leave a reference to the joined SessCoP instead.

An alternative is the introduction of an independent and dedicated node to maintain the session configuration profiles. However, there is communication necessary for each access, which impacts the performance of such a solution.

A distribution of the information to the SSP would bring advantages for the actuality of the SessCoPs' information, as the actual state of the participants could be accessed easier, but the communication overhead due to the distributed nature of the information on the relatively slow signalling network would be significant. However, this restriction does not hold for an IN architecture based on the B-ISDN.

Performance. Service processing in the IN is a time critical task, and it can be taken for granted that the proposed methodology influences this service processing time. There is a direct relationship between the performance and the delays that are introduced by the access and processing of the service configuration profiles as well as an indirect relationship caused by the higher complexity of the service logics. These important issues need more investigation in further research.

5 Related Work

In contrast to the application of a *Feature Interaction Manager (FIM)* as presented in [9], in the SAC approach presented here the knowledge about the resolution of the conflicts between the services is not concentrated to one functional entity – the FIM – but distributed in a generalized form to the services. The *table driven approaches* described by Homayoon and Singh in [10] and by Schessel in [11] decide in the SSP on the activation of features based on static decision tables, in a feature individual way.

In the approaches described above, only interactions of services triggered in the same call state model can be handled properly, while in the SAC approach presented here these interactions as well as interactions between services triggered in different SSPs, but in the context of the same session, are addressed, as the decisions are made during service processing, not in the SSP.

Griffeth's and Velthuijsen's *negotiating agents* approach [12] is based on a negotiation process based on a general hierarchy of goals common to all services. One service makes a proposal how to proceed based on its own goals and preferences. Another service checks whether this proposal can be accepted with respect to its own goals, or whether a counterproposal has to be made. Although this approach aims at interactions on the logical level, there are some structural similarities to the SAC approach in a sense that there is a service independent representation of the results of the services' actions on a session. However, in the SAC approach, no abstract goals representing intentions for the future, but only concrete facts representing actions that have taken place in the past, are available. Thus, no negotiation process takes place, but only a simple decision based on the SessCoP data. It cannot be guaranteed, that interactions that manifest itself as logical contradictions in the services' intentions are detected and resolved.

6 Conclusions

The results and proposals presented in this contribution result from studies of service and feature interactions in Intelligent Networks. The significant progress that has been achieved in the area of detection of feature interactions during the last years has not been complemented by comparable success in the resolution and prevention of these situations. The reason for this fact is, in the author's opinion, twofold. It is relatively easy to detect logical and intentional contradictions between services provided there is a suitable notation for these issues, but it is difficult to resolve them, like in the negotiating agents approach. But even if these problems can be solved, there remains a significant number of interaction cases, where it is clear how the services should react in principle, but it cannot be done or it is done in a wrong way as it is not enough information available to carry out the task correctly.

In this paper, we focus on the latter type of interactions. The concept of *Service Admission Control* is introduced and a methodology to resolve and prevent unwanted interactions is presented that is based on a dynamic *Session Configuration Profile* for each session that is influenced by at least one IN service. This profile can be accessed and modified by all IN services participating in a session. They can gather all the information needed to react properly that is not available in the ISUP and INAP signalling messages, provided that this information is contained in the session configuration profile's data model, and they can use resource reservation mechanisms that are not supported by the underlying network.

As this contribution reports on work in progress, there are many open issues for further research. There is at the moment only a vague idea of the necessary contents of the Session Configuration Profile, as well as of the methods needed to access and manipulate these contents. Furthermore, it has to be investigated to what degree the service design will be complicated by taking into account the maintenance of a SessCoP. On the side of non-functional properties, all the allocation alternatives and the performance impact of the proposal have to be studied thoroughly.

References

1. T. Magedanz and R. Popescu-Zeletin, *Intelligent Networks: Basic Technology, Standards and Evolution*, International Thomson Computer Press, 1996.
2. I. Faynberg, L. R. Gabuzda, M. P. Kaplan, and N. J. Shah, *The Intelligent Network Standards: their application to services*, McGraw-Hill, 1997.
3. E. J. Cameron and H. Velthuijsen, "Feature Interactions in Telecommunications Systems", *IEEE Communications Magazine*, vol. 31, no. 8, pp. 18–23, August 1993.
4. N. D. Griffeth and Y. J. Lin, "Extending Telecommunications Systems: The Feature-Interaction Problem", *IEEE Computer*, vol. 26, no. 8, pp. 14–18, August 1993.
5. E. J. Cameron, N. D. Griffeth, Y. J. Lin, M. E. Nilson, W. K. Schnure, and H. Velthuijsen, "A Feature Interaction Benchmark for IN and Beyond", in *Feature Interactions in Telecommunications Systems*, L. G. Bouma and H. Velthuijsen, Eds., May 1994, pp. 1–23.
6. P. Combes, M. Michel, and B. Renard, "Formal Verification of Telecommunication Service Interactions using SDL Methods and Tools", *Proceedings of the 6th SDL Forum 1993 (SDL '93)*, pp. 441–452, 1993.
7. "Principles of Intelligent Network Architecture", ITU-T Recommendation Q.1201, Geneva, 1992.
8. P. K. Au and J. M. Atlee, "Evaluation of a State-Based Model of Feature Interactions", *Feature Interactions in Telecommunication Networks IV*, pp. 153–167, June 1997.
9. M. Cain, "Managing Run-Time Interactions Between Call-Processing Features", *IEEE Communications Magazine*, vol. 30, no. 2, pp. 44–50, February 1992.
10. S. Homayoon and H. Singh, "Methods of Addressing the Interactions of Intelligent Network Services With Embedded Switch Services", *IEEE Communications Magazine*, vol. 26, no. 12, pp. 42–70, December 1988.
11. L. Schessel, "Administrable Feature Interaction Concept", *Proceedings of the XIV International Switching Symposium (ISS)*, pp. B6.3–B6.3, October 1992.
12. N. D. Griffeth and H. Velthuijsen, "The Negotiating Agents Approach to Runtime Feature Interaction Resolution", in *Feature Interactions in Telecommunications Systems*, L. G. Bouma and H. Velthuijsen, Eds., May 1994, pp. 217–235.

INAP Protocol Test Suite Verification Method Using the IUT Simulator for AIN System Conformance Testing

Hyunsook Do, Seongyong Bae, Sangki Kim

Intelligent Network Architecture Section
Electronics and Telecommunications Research Institute
YUSONG P.O.BOX 106, TAEJON, 305-600, KOREA
E-mail: (hsdo, sybae, sangkim)@dooly.etri.re.kr

Abstract The INAP(Intelligent Network Application Protocol) protocol test suite is the test procedures to conform that the implemented INAP protocol is equivalent to the standard specification of the INAP. The INAP test suite is generated automatically by using formal description technique. Before the generated test suite is applied to the real target system, it should be verified that this test suite has been correctly built according to the INAP specification. In this paper, we implemented the IUT(Implementation Under Test) simulator on the K1197 tester to verify the generated INAP test suite. This method enabled us to correct the syntactic errors in the test suite and unexpected test sequence errors. It guarantees the reliability for verified results and reduces time and costs to verify the test suite than the manual method based on the protocol specification represented by natural language.

1 Introduction

The growing demand for data communications has led to the standardization of several communication protocols which are implemented in many different systems by different vendors[5]. Conformance testing is to ensure their interoperability these implementations. To detect errors, the conformance tester exchanges messages with the IUT according to a test scenario.

A test suite, a set of test scenarios, is usually generated from the protocol specification manually or automatically using formal description technique. In this paper, we adopt the automatic test suite generation method because this method can correctly describe protocol specification and reduce time and costs.

Before the generated test suite is applied to the real target system, it is verified to prove that this test suite has been correctly built according to the INAP specification. In this paper, we implemented the IUT simulator on the K1197 tester which has the CCS7(Common Channel Signalling System No.7) environment to verify the generated INAP test suite. This method enabled us to correct the syntactic errors and unexpected test sequence errors in the test suite. It guarantees the reliability for verified results and reduces time and costs to verify the test suite than the manual method based on the protocol specification represented by natural language. And also the IUT simulator can be reused to verify the modified or extended test suite according to the INAP evolution toward CS-2/3(Capability Set) with little efforts

In the next section, we describe the testing method and the test configuration for the INAP testing. Section 3 explains the INAP suite generation processes and section 4 describes the IUT simulator's implementation processes and its function. Section 5 mentions about the test suite verification using the IUT simulator and practical experiences. Finally, section 6 discusses conclusions.

2 Testing Method and Test Configuration

2.1 Test Method

IS 9646 describes four Abstract Test Methods: Local Test Method, Distributed Test Method, Coordinated Test Method, Remote Test Method[2]. These methods can be properly chosen by test environment and test purpose. For the INAP conformance testing, the remote test method is selected primarily due to its minimal impact on an IUT. It is not possible in this test method to observe and control the upper service boundary of the IUT.

The abstract model for the Remote Test Method is shown in Figure 1. The LT(Lower Tester) is a test component communicating with the IUT through PCO(Point of Control and Observation) at the IUT lower interface, and the UT(Upper Tester) is a test component communicating with the IUT through PCO at the IUT upper interface.

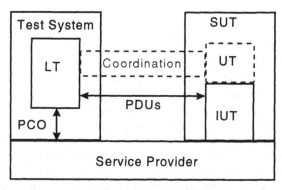

Figure 1. Remote Test Method

2.2 Test Configuration

The test configuration for the SSP(Service Switching Point) INAP test is presented in figure 2[9]. The lower protocol stack of the INAP is omitted in this configuration. Three LTs are defined. The LT1 is a main LT which simulates the SCP(Service Control Point) and uses the INAP interface. The PCO1, as a main PCO, exists in the interface between the LT1 and the SSP INAP IUT. The LT2 and LT3 are used to simulate the function of originating/terminating terminals for testing special INAP procedures and connected to the SSP with non-INAP interface, DSS1(Digital Subscriber Signalling System No.1). The PCO2 and PCO3 are placed in the interfaces of IUT-LT2 and IUT-LT3, respectively. Figure 3 and 4 show the test configurations for the IP(Intelligent Peripherial) and SCP INAP test[9].

The LT4 and LT6 are main LTs simulating the SCP and SSP, respectively. They are connected with the INAP interface. The PCO4 and PCO6, as main PCOs, exist in the LT4 - IP INAP interface and LT6 - SCP INAP interface, respectively. The LT5 is used to simulate the SSP for testing special INAP procedure and connected to the IP with non-INAP interface, DSS1. The LT7 is used to simulate the IP for testing special INAP procedure and connected to the IP with the INAP interface. The PCO5 and PCO7 are placed in the IUT-LT5 interface and the IUT-LT7 interface, respectively. Each CP(Coordination Point) in the all test configurations is used to exchange coordination messages between LTs.

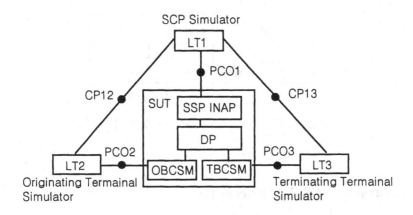

Figure 2. Testing configuration for SSP INAP test

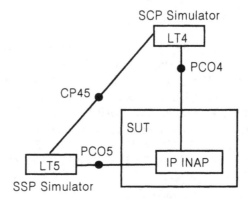

Figure 3. Testing configuration for IP INAP test

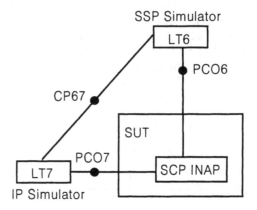

Figure 4. Testing configuration for SCP INAP test

3   Abstract Test Suite Generation

Figure 5 shows test suite generation processes from INAP specification by FDT(Formal Description Technique) method. This figure describes the test behaviour generation part and data description part. We used Verilog's ObjectGEODE as a FDT tool[14,15,16,17,18] and ITEX as a TTCN(Tree and Tabular Combined Notation) editor[19]. ObjectGEODE provides SDL(Specification and Description Language) editor, MSC(Message Sequence Chart) editor , simulator and TTCN generator. The TTCN generator converts SDL and MSC descriptions into test sequence in TTCN. ITEX provides TTCN editor, analyzer and GR(Graphical Representation) postscript. ObjectGEODE can handle the data part of ATS(Abstract Test Suite) partially, but it can't support ASN.1 notation which is the standard for INAP data description. For this reason, we use ObjectGEODE only for generating of test sequence, while we define the data description of ATS manually using ITEX.

The FSM(Finite State Machine) for SSF is modeled by SDL language from INAP specification recommended by ITU-T. Test purposes are also described using formal language, MSC. The FSM model and test purposes are verified and validated to detect errors of the system by the exhaustive simulation performed on the SDL and MSC model. The constrained state graph of the SDL model is generated through the exhaustive simulation. TTCN generator produces test sequence from constrained state graph.

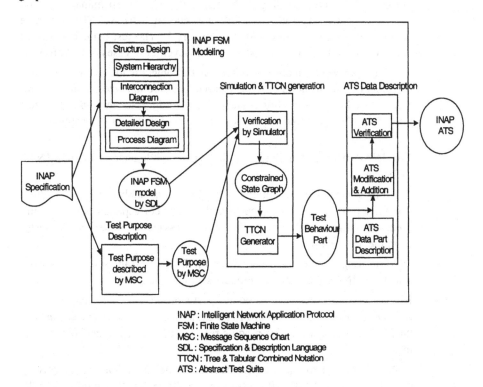

Figure 5. Abstract Test Suite generation processes

4   IUT simulator implementation

To verify the generated test suite, we implemented the IUT simulator on the K1197 tester. The K1197 tester has KT(Korea-Telecom)-INAP user part and CCS7 signalling link to operate the IUT simulator. The KT-INAP user part builds the test messages complying with the KT-INAP specification using message building system. The IUT simulator has a message pool and two functional blocks: simulation functional block and testing data management functional block. The message pool has data, which are used in the functional blocks. The functional blocks can retrieve and modify data in the message pool and also create data from the INAP specification. The IUT simulator functional blocks is shown in Figure 6 and each functional block has several modules and their functions are followings.

- Simulation functional block
  - Test case identification module : Conformance testing is executed a unit of test group or a test case by the test purposes. This module identifies the object to be tested according to the test unit from the test manager. To do this, it uses the test suite structure information database. The test suite structure has tree structure shown in Figure 7. The test suite is made of several test groups and each test group has several test cases or test groups.
  - Test case procedure analysis module : If test object is identified in the test case identification module, its test procedures are analyzed using the test case procedure information database. According to the analyzed test procedure, the test messages are sent to the IUT simulator or received from it.
  - Non-INAP message handling module : This module handles the non-INAP messages.
  - INAP message selection module: According to the analyzed test procedure, if the test procedure to be executed at this stage corresponds to the sending procedure, this module selects the INAP message to be sent to the tester from the message pool.
  - INAP message sending module : After the INAP message selection module selects the INAP message this module sent it to the tester.
  - INAP message receiving module : According to the analyzed test procedure, if the test procedure to be executed at this stage corresponds to the receiving procedure, this module receives the test message from the test manager.
  - INAP message error detection module : If the INAP message receiving module receives the INAP message from the tester, this module checks whether the message has errors or not.

- Testing data management functional block
  - Testing data creation module : This module inputs the INAP specification information into the message pool.
  - Testing data modification & deletion module : The test message can be modified or deleted by user.

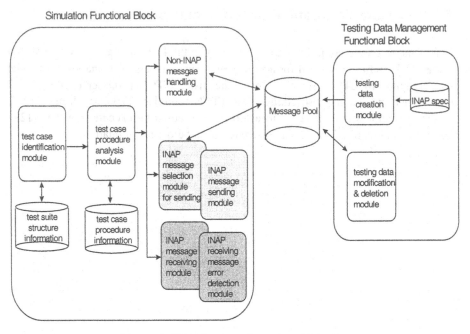

Figure 6. IUT Simulator Functional Blocks

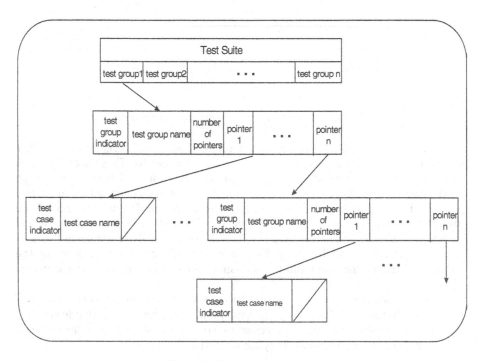

Figure 7. Test Suite structure

5   Test Suite Verification using IUT simulator

Before the test suite could be used in testing with the IUT simulator it was analyzed using ITEX and the reviewed for accuracy and completeness. The analyzed test suite was compiled to ETS(Executable Test Suite) using TTCN compiler on the K1297 tester. There were some restrictions of the TTCN compiler supplied with the K1297, so the test suite was modified into a form which could be executed on the K1297 tester. Errors identified at this stage were removed before recompilation.

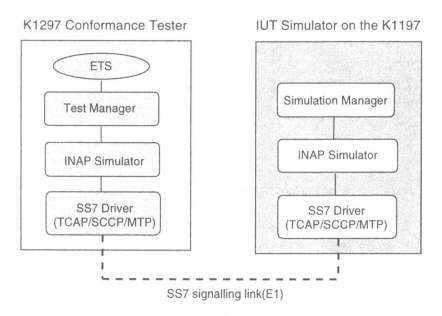

Figure 8 The INAP test suite verification environment

Figure 8 shows the INAP test suite verification environment using the IUT simulator. The IUT simulator was run on the K1197 which has CCS7 signalling link environment. The tester and the IUT simulator communicate with each other through the E1 interface. In the K1297 tester, the running test suite is connected by an internal software interface to the run-time operation system. The interface was comprised of a user-defined test manager and INAP simulator. The test manager performs the test execution with ETS and the test log data management. The INAP simulator provides the INAP specification table called INAP user part, which is used for decoding test messages between the tester and the IUT simulator so that the user can understand them.

Figure 9 describes the test suite verification processes on the K1197(the IUT simulator). Because the IUT simulator displays the error messages during testing, we can correct it easily. During this processes we could find some errors in the test suite and K1297 dependent problems. They are summarized followings.

- The test suite has redundancy for defining the return error handling

- Some test cases did not assign verdicts
- Some test cases have wrong test steps, especially TCAP(Transaction Capabilities Application Part) primitives concerned with transaction handling
- Some data in the test suite were not complied with the INAP specification
- Some K1297 dependent problems concerned with non-INAP messages handling were not solved. This problems require the K1297 software enhancement

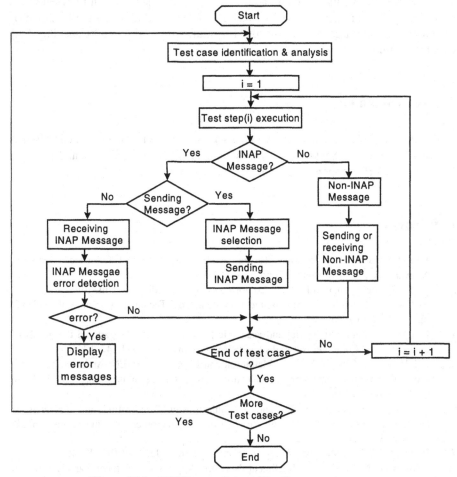

Figure 9 The INAP test suite verification processes

The problems issued above were solved except the non-INAP message handling. This unsolved problem requires the K1297 tester's enhancement. For the time being, we could handle the non-INAP messages manually.

6 Conclusions

In this paper, we described the test suite verification method by implementing the IUT simulator on the K1197 tester. The IUT simulator was used to prove that the generated test suite has been correctly built according to the INAP specification. We could find and correct the test suite errors through this process. This method enabled us to correct the syntactic errors in the test suite and unexpected test sequence errors. It guarantees the reliability for verified results and reduces time and costs to verify the test suite than the manual method based on the protocol specification represented by natural language. Further, the IUT simulator can be reused to verify the modified or extended test suite according to the INAP evolution toward CS-2/3(Capability Set) with little efforts

Acknowledgement

The author would like to thank Dr. Kyung Pyo Jun, and the members of the Intelligent Network Architecture Section for their valuable advice and collaboration. This work is a part of the project sponsored by the Korea Telecom(KT).

References

1. Behcet Sarikaya, "Conformance Testing: Architectures and Test Sequence," Computer Networks and ISDN Systems 17, pp.111-126, 1989.
2. Keith G. Knightson, OSI Protocol Conformance Testing, McGraw-Hill, 1993.
3. M C Bale, "Signaling in the intelligent network," BT Technical Journal, Vol.13, No.2, pp.30-42, April 1995.
4. R. Lai, W. Leung, "Industrial and academic protocol testing: the gap and the means of convergence," Computer Networks and ISDN Systems 27, pp.537-547, 1995.
5. S T Vuong, H Janssen, Y Lu, C Mathieson and B Do, "TESTGEN: An environment for protocol test suite generation and selection," computer communications, Vol.17, No.4, pp.257-270, April 1994.
6. T Ramalingam, Anindya Das and K Thulasiraman, "On testing and diagnosis of communication protocols based on the FSM model," computer communications, Vol.18, No.5, pp.329-337, May 1995.
7. ETSI, IN; Global functional plan for IN Capability Set 1, TCR-TR 016, 1994.
8. ETSI, IN; IN CS 1 Core Intelligent Network Application Protocol Part 1: Protocol specification, ETS 300 374-1, 1994.
9. ETSI, IN; IN CS-1 Core INAP Part 4: ATS and PXIXIT Proforma Specification for SSF and SRF, ETS 300 374-4, 1995.
10. ITU-T Revised Recommendations, Interface Recommendation for IN CS-1, Q.1218, 1995.
11. ITU-T Baseline Document for IN CS-2, 1995.
12. ITU-T SG11 WP4 TD PL/11-11,12 Q.1214, "Distributed Functional Plan for IN CS-1," April 1995.

13. ITU-T Recommendation Q.1201, "Principles of Intelligent Network Architecture," 1992.
14. An Introduction to SDL and MSC, Verilog, 1995.
15. ObjectGEODE Method Guideline, Verilog, 1995.
16. GEODE Editor Reference Manual, Verilog, 1995.
17. ObjectGEODE Simulator, Verilog, 1995.
18. ObjectGEODE TTCgeN User's Guide, Verilog, 1995
19. ITEX User's Guide, Telelogic, 1994
20. K1297 Protocol Tester Basic Manual, Siemens, 1996

Testing IN Protocol Implementation

P. Florschütz

Siemens Nixdorf Informationssysteme AG

Abstract. Testing correctness of protocol implementation of an IN platform is a demanding task. In this paper we would like to give a short outline of our *SSPSimulator* protocol testing software, compare it to our *K1197* hardware protocol tester and discuss differences and use of it.

1 Notation

Within this document we use certain font conventions: In narrative text anything related to **INAP** protocol is printed in bold capital sans serif letters, for TCAP, SCCP and MTP levels we use unbold capital sans serif letters. Keywords, scripts and scenario examples are printed in `typewriter font`. When giving short examples in normal text we use *slanted fonts*.

2 Basics

2.1 Tasks of Testing

Several aspects may be considered:

- General requirement: diagnosis of errors in **INAP** implementation;
- Functional tests with/without consideration of time constraints;
- Load tests, regression tests, net simulation (parallel execution, multiple SSP simulation, error simulation, statistical features, delays);
- Tests for system durability against external protocol offenders (accidental drop of protocol messages);
- Reproduction of errors for debugging purposes.

2.2 Intended Quality of Testing

The testing tools being mentioned in this paper are mostly used for software development and debugging purposes. They do not replace testing the whole system in a real heterogeneous physical net environment, but they are strong enough to check functional quality of new code or changes in old code. On the other hand these tools are used to create quite specific scenarios to reproduce or just analyse erraneous behaviour on protocol level which helps drawing conclusions for several components of the IN system. These are our main issues in testing:

- general requirement: maximum testing efficiency;
- ability to generate nearly any test case needed for testing some implementation;
- reduce time consumption of testers;
- in a way easy to use, no need to be a protocol expert, but in fact, some knowhow is necessary, hide nasty details;
- delivery of diagnosis files in a convenient way;
- non interactive use possible;
- in the long term tools need to be extensible to keep up with protocol enhancements.

2.3 Simple Test Scenarios

Consider these examples, that are subject to frequent test sessions:

- Successful sending of **INAP** operation:
  *Sending **CIQ** with **callConnectedElapsedTime** requested before **CON** should for instance result in some **CIR** giving call duration as soon as call is finished. The data obtained may be used for statistical evaluation.*
- Successful receipt of **INAP** operation:
  *Receipt of **SFR** operation should initiate corresponding increase of votum counters.*
- Timeout of **INAP** operation:
  *Timeout of **AT** operation (30min timer) should result in forced finishing dialogue by sending TC_U_ABORT*

2.4 Software and Hardware Simulation

While our *SSPSimulator* software allows multiple invocation, parallel execution and wide ranged test load scalability (by means of underlaying test driver machine's efficiency) it is subject to generic software simulation problems like interference with tested software, falsified test constraints (e.g. realtime constraints offended due to heavy simulator load) and others. *SSPSimulator* is platform dependent as it uses a certain communication software including SS7 protocol stack and transaction capabilities (**TCAP**) layer and therefore cannot be used as a generic protocol simulator. On the other hand it allows automatic execution of highly complex collections of test scenarios and generates detailled test protocols to be reviewed in case of test failure. Also, it allows statistical distribution of distinct answer messages, accidental message dropping etc. *SSPSimulator* can take control over time scale and thereby may reduce testing time in some case, as SCP idle periods within calls may be skipped by increasing time. Note that this does not influence running **INAP** operation timer and other SCP timer. This means it is not quite applicable as long as there are open calls.

Hardware simulation by use of a generic protocol tester like our *K1197* provides a more realistic test environment as it is simply plugged in and acts physically equal to an ordinary SSP. Also –in our case– it grants more possibilities to parameterize message than our *SSPSimulator* does. It can easily be used to check for problems at customer sites where it may not be possible to install a complete software testing environment (test driver machine or, at least, software simulation running on IN system itself). It is a generic device that may be used independent of any IN protocol software. On the other hand the test device requires special hardware, software, configuration and maintenance knowledge and may not be suitable for load tests and running complex *hypertestplans* (test plans consisting of several defined stages and script collections with automatic scanning of protocol and reports).

3 Testing with *SSPSimulator*

3.1 Complete System in Test

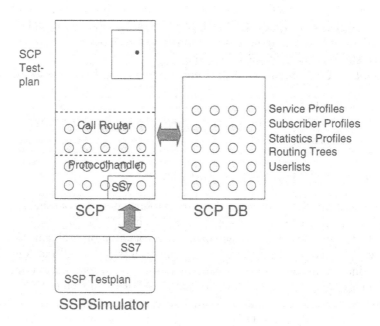

Fig. 1. System under test

Obviously it is not enough to just send an **IDP** message to SCP. Fig. 1 shows a very simple minimal test environment, that must be maintained and initialized for successful testing. Only the protocol data is generated by *SSPSimulator*. To make the SCP do something reasonable it needs several sources of information apart from it's protocol handler. This means that any SCP test plan will spool in service- and userdata and routing trees (that need to be tailored and administrated individually) before starting the protocol simulation. In fact, such a preinitialization always is part of our test plans, but is not within scope of this document as it is strongly related to our IN architecture.

3.2 Functional Scope of *SSPSimulator*

SSPSimulator is characterized by certain issues:

- protocol level: **SINAP3** (extended) operations; TCAP messaging; adjustable SCCP level addressing;
- ASCII notation for test scripts;
- support for active and passive dialogues (send/receive first message);
- allowance of incomplete component specification;
- multiple execution of test scripts with statistical variation of message type;
- programmable state machine triggered by incoming events;
- optional verification of responses (complete/partial);
- optional simulation of message loss;
- configurable test report, load chart and error statistics generation.

3.3 Design of *SSPSimulator*

SSPSimulator is designed as well configurable IN test tool suitable for distributed system architecture.

How It Works Generally speaking, *SSPSimulator* simulates a TCAP level (cf. sec.3.4) communication partner that is capable of a certain **INAP** variety. This also means it may be used to simulate both ends of communication.

Finite state machines (FSM) are commonly used to implement protocol handlers for switches and IN systems as well. *SSPSimulator* does not simulate a complete switch protocol FSM in itself (there is no inherent switch intelligence, as it is only a simulation), but interacts by interpretation of ASCII test scripts that allow specification of an event driven protocol state machine by telling *SSP-Simulator* what to send and what to expect. However, *SSPSimulator* keeps track of the current state by maintaining a *call context* (cf. sec.3.3) for each *dialogue* (cf. sec.3.4). This means there is no knowledge about dialogue structures coded into *SSPSimulator*. On the other hand, the tool knows how to encode valid **INAP** operations and can automatically supply an operation's parameters as far as they are missing in the test specification such that a test script may focus on test related parameters and does not need to put up with irrelevant details.

Call Handling A simulated call is represented by a unique *call context* within *SSPSimulator*. Message flow control is implemented by a call FSM. The *call context* implements the abstract state of the call FSM in terms of **TCAP/INAP** transactions within one peer. Each call context is assigned a unique *context ID* that is also used as *originating transaction ID* (cf. sec.3.4) for the related dialogue when sending a message to the SCP. The latter –when receiving that message–

calculates it's own transaction ID for this dialogue and sends it to *SSPSimulator* by using it as originating transaction ID (which then becomes the destination transaction ID for *SSPSimulator*). The *SSPSimulator* context manager uses the *SSPSimulator* transaction ID as search key for the corresponding call context.

Interface to IN Platform *SSPSimulator* needs an external SS7 protocol stack implementation (TCAP – SCCP – MTP) that it uses to send it's TCAP messages to the IN application to be tested. It is a matter of SCCP addressing to run *SSPSimulator* on same node as test system or run it on a remote node. This does not make any difference to *SSPSimulator*. However, system load is to be taken into account, as *SSPSimulator* requires quite some load capacity to fulfil it's supervision and test driver capabilities. For performance and regression test issues we mostly use distinct driver machines, for development purposes an the other hand, it is perfectly valid to run *SSPSimulator* on test system itself.

3.4 Protocol Implementation

Some TCAP Basics TCAP (*Transaction Capabilities Application Part*) is an IN application level protocol grounded on SS7 protocol suite. In a generic way it allows invocation of operations on remote communicating entity combined with (optional) feedback messages. These operations are enclosed and organized in *components* accompanied by TCAP *component handling primitives* (CHP) each, that in itself are part of and organized in TCAP *dialogues* that consist of TCAP *dialogue handling primitives* (DHP). Hence we have dialogues (formed by DHPs) that may each contain one or more components (denoted by CHPs) that –finally– carry our operations.

Any operation usually is assigned an operation timer that –when expired– may cause (implementation dependent) certain protocol actions like negative acknowledge (or even no action at all). *SSPSimulator* uses TCAP for sending and receiving ASN.1 coded **INAP** (*IN Application Part*) operations according to our **SINAP3** standard. As a matter of fact it therefore is capable of ASN.1 encoding and decoding and knows about structure and syntax of **SINAP3** operations.

As mentioned, information exchange is organized in TCAP dialogues that are distinguished by assigning a unique originating or destination transaction ID, depending on referring to the initiator's or partner's view.

Dialogue Modelling *Dialogues* in our terms are (in a protocol related manner) closed entities of communication between SSP/simulator and SCP (cf. sec.3.4). A dialogue may be *call associated* (CA) or *non-call-associated* (NCA), depending whether its invocation is triggered by a phone call or by other net action (like call gapping). We also distinguish between *active* or *passive* dialogues, where per definition the sooner is invoked by sending while the latter is triggered by receipt of a message.

For *SSPSimulator* any dialogue is represented by a *call context* (cf. sec.3.3) that uniquely is created for and related to this specific dialogue. For each dialogue

SSPSimulator obviously has to keep at least two states, a *sending* and a *receiving* state of the dialogue. Meaning of state transitions:

– state change of receiving state by
  • receipt of a message;
  • timeout of message receipt;
– state change of sending state by
  • sending of a message;
  • simulated loss of a message to be sent;

Any dialogue's components consist of TCAP primitives carrying **INAP** operations. **INAP** operations are divided into *synchronizing* and *non synchronizing operations*, depending on their quality of triggering a state transition within sending and receiving entity. E.g. **FCI** and **SCI** are non synchronizing while **CON** is a synchronizing operation as sender changes it's dialogue state on sending.

Specification of *SSPSimulator* Messages *SSPSimulator* defines and uses some sort of a limited set of "message templates" that are used to specify the TCAP messages to be sent. These message templates are called *SSPSimulator events* and normally do specify one TCAP DHP with it's components (optional). *SSPSimulator* events mostly look like e.g.

EVENT_DISCONNECT_B ::= Cnt [ACR] [CIR] ERB

where Cnt denotes the TC_CONTINUE DHP, **ACR** and **CIR** denote optional components to be included in resulting message and **ERB** denotes the last and mandatory component. *SSPSimulator* will send a TC_CONTINUE DHP with 1 to 3 components (according to test script). This is an example from a valid test script:

```
MESSAGE EVENT_DISCONNECT_B ERB_Disc_B =
  {
    ACR_SEQ_INFO            = FINAL;      // final
    ACR_USED_TIME           = 120;
    ACR_USED_TIME_AFTER_SW = 0;           // no TimeSwitch
  };
```

This makes *SSPSimulator* issue a TC_CONTINUE DHP with a final **ACR** and an **ERB** component reporting a disconnect B event.

Additionally it is possible to specify any hexadecimal TCAP message. *SSPSimulator* then will send it basically as is without doing any further encoding. Thus it will replace destination transaction ID, for instance. This is an example for a hexadecimal TC_U_ABORT:

```
MESSAGE U_ABORT U_ABORT_cause16 =
{
  25                        // length of TCAP message
  67                        // U_ABORT tag
  23                        // length
  49                        // tag for dest trans id
  04                        // length dest trans id
  04 1c 00 01               // bytes 29-32: Dest Trans ID, any value will do
```

```
6b                          // dialog portion tag
1b                          // length og dialog portion
28                          // tag (dialog external)
19                          // length
06                          // tag AS id
08                          // length of AS id
0a 82 06 01 05 00 02 00    // AS id
a0                          // tag UserInfoSingleAsn1
0d                          // length
30                          // tag S3-userAbortInfo
0b                          // length
a0                          // tag UserAbortCause
05                          // length
a0                          // tag AbortedBySSP
03                          // length
80                          // tag networkEvent
01                          // length
02                          // value
81                          // tag UabortCause
02                          // length
80                          // val   (coded into 2 bytes)
90                          // val
};
```

Collection of Supported Transactions Here we point out some examples
for valid *SSPSimulator* message types. This is far from being complete.

Application Context An application context may be specified with any **TC_BEGIN**.
Example:

```
MESSAGE INITIAL_DP   IDP_179-101-123 =
  {
    APPL_CONTEXT_NAME = 02 81 68 01 01 01 01; // dialog AS ID
    IDP_CGPA  = 1-2141201, 03 11;
    IDP_CDPA  = 179-101-123, 83 10;
  };
```

where **INITIAL_DP** per definition denotes a **TC_BEGIN** with an optional Dialogue Portion and an **IDP** operation. APPL_CONTEXT_NAME specifies the application context name and IDP_CGPA and IDP_CDPA denote calling and called party
(digit) with odd/even indicator, nature of address indicator, internal network
number indicator and numbering plan indicator added in two trailing hex digits
as binary representation.

Call Events Specification of call events is possible like so:

$$\text{EVENT\_BUSY} ::= \text{Cnt [ACR] [CIR] ERB}$$

Example:

```
MESSAGE EVENT_BUSY ERB_net =
{
  ERB_BUSY_CAUSE = 82 af;  // 10000010 10101111 (47: res. unavail.)
  // B\b   8       7        6       5       4       3       2       1
  // --+-------+---------------+-------+-------------------------------
  // 1 |   1   | coding stand. | spare |         location
  // --+-------+---------------+-------+-------------------------------
  // 2 |   1   |                 cause value    (7 bit)
  // --+-------+-----------------------------------------------------
```

Here we do not send **ACR** and **CIR**. This results in an **ERB** being sent with
busy cause 47.

Charging SSPSimulator supports online charging by means of **AC** and **ACR** operations and net charging operations (**FCI**, **SCI**) as well.

NCA Operations It is possible to generate and receive **ASF** (Activate Service Filtering), **SFR** (Service Filtering Response) and **CG** (Call Gap) operations.

3.5 Example

Here we show an example where *SSPSimulator* simulates a GSM switch communicating with an IN Service Control Point. This would be a typical test situation for us.

Scenario The example shows a simple case where an incoming call is indicated by a regular **IDP** (Initial DP) message. Caller is 030-3132575, called party is 179-341-12355534 which is a GSM IN number to be routed by SCP. The SCP then is expected to send **CON** (Connect) to a destination number 55534 (that has to be obtained by the SCP's IN capability) together with an **AC** (Apply Charging) operation. Heart beat time is set to 2 min. The SCP also supplies e parameters to be sent to mobile phones for cost information (some providers do not like this being transmitted to the phone). Granted time (overall speechtime until call end or tariff switch) should be calculated by SCP and result in 805 sec. After 1sec *SSPSimulator* sends **ERB** (Event Report BCSM) with Disconnect B (B party has dropped phone) event together with a final **ACR** (Apply Charging Report) after 1 sec of call time). SCP then should automatically send an announcement of remaining credit and release call, all coded into one **TCAP** message specified by an END_ANNOUNCEM *SSPSimulator* event.

Additional Initialization According to sec. 3.1 we need to prepare the SCP and it's database. The actual SCP test plan calls some additional tools that do database initialization with realistic SCP data. In this example we might have an extremely simple GSM VPN service, that might be used for special charging on certain numbers. A number is dialled by use of service 341 which is routed and charged by IN criteria. The SCP might check calling and called party address, call time, caller credit, etc. Respective information must be placed in DB (see fig. 1) before test may be started. We do not discuss the test plan here, that calls the following test script.

Test Script

```
TRLIMIT = 1;
DISPLAY = 3;

// Message format: TC_BEGIN IDP
MESSAGE INITIAL_DP    IDP_179-341-12355534 =
    {
    IDP_CDPA  = 179-341-12355534, 03 10;
    IDP_CGPA  = 030-3132575, 03 11;
    };
```

```
// Message format: TC_CONTINUE [AC] [DFC] [CIQ] [RRB] [FCI [SCI]] CON
MESSAGE CONNECT CN_55534 =
    {
    CON_DEST      = 55534, 81 10;
    AC_TIME_GRANTED = 814; // granted time (calculated by credit)
    AC_WBR_TIME   = 30;  // warning tone 30 sec before release
    AC_WBR_TONE_ID = 1;   // warning tone id
    AC_HEART_BEAT = 2;   // heartbeat timer (min)
    AC_E1         = 10;  // set of e parameters
    AC_E2         = 100;
    AC_E3         = 100;
    AC_E4         = 200;
    AC_E5         = 0;
    AC_E6         = 1;
    AC_E7         = 150;
    SCI_DATA   =;          // not relevant
    };

// Message format: TC_CONTINUE [ACR] [CIR] ERB
MESSAGE EVENT_DISCONNECT_B ERB_Disc_B =
  {
    ACR_SEQ_INFO        = FINAL;  // final heartbeat comes
    ACR_USED_TIME       = 1;      // together with ERB
  };

// Message format: TC_END [DFC] [FCI [SCI]] [CTR] PA RC
MESSAGE END_ANNOUNCEM Announcem_account =
    {
    ITS_MSGID_TYP  = 1034;
    ITS_VM_PRICE   = 00 00 08 00;
    };

// message flow
DIALOG Dlg_1 =
    {
    send    (IDP_179-341-12355534 ); // Initial DP -->
    receive (CN_55534);              // Connect to destination <--
    send    (ERB_Disc_B);            // ACR with ERB -->
    receive (Announcem_account);     // Announcement of credit <--
    };

// start dialogue
EXECUTE
  {
    Dlg_1;
  };
```

Testprotocol/Message Flow

```
This protocol was created by SSPSIM (Vers. 28.Mai.96)!
By: A. Krenn, PSE TMN 34, Wien-GUD, Sep 26 1997
Executing Testplan SSPBSP
Writing Protocol to /home2/pf/w/SSPBSP.1
Date/Time of Creation: Fri Oct 17 14:53:54 1997

OWNTASK=, DEST=, TPFILE=SSPBSP
DELAY=10 ms., TRLIMIT=-1, TOUT=20 s
Testplan: SSPBSP

starting dialog 'Dlg_1' with trid=1000

  0.01: INITIAL_DP(IDP_179-341-12355534) ========> sent to
IDP_CDPA       = 17934112355534f0, 83 10
IDP_CGPA       = 0303132575, 03 11
---------------------------------------------------
```

```
  0.11: <======= CONNECT(CN_55534) from SE_MAIN:
RRB_EVENTS       = DISCONNECT
CON_DEST         = 555340, 81 10
AC_TIME_GRANTED  = 814
AC_WBR_TIME      = 30
AC_WBR_TONE_ID   = 1
AC_HEART_BEAT    = 2
AC_E1            = 10
AC_E2            = 100
AC_E3            = 100
AC_E4            = 200
AC_E5            = 0
AC_E6            = 1
AC_E7            = 150
-------------------------------------------------

  0.11: EVENT_DISCONNECT_B(ERB_Disc_B) =======> sent to
ACR_SEQ_INFO     = FINAL
ACR_USED_TIME    = 1
-------------------------------------------------

  0.17: <======= END_ANNOUNCEM(Announcem_account) from SE_MAIN:
FCI_DATA         = x 80 ff 00 01 01 01 00 00 00 06 31 37 39 33 34 31
05 00 00 01
CTR_IP_DEST      = 9204, 03 10
ITS_MSGID_TYP    = 1034
ITS_VM_PRICE     = 00 00 08 00
ITS_NOR          = 1
ITS_DURATION     = 60
PA_DISC_IP_FORB  = NO
PA_RA_CMPL       = NO
RC_REL_CAUSE     = x 82 80
-------------------------------------------------

-DIALOG 'Dlg_1'--------------------------------------------
|         SSPSIM         t:1000  | DEST=        t:1      |
s    t    ------------------------|-----------------------m--|
0   0.01  INITIAL_DP IDP_179-341-12355534---------------------->
1   0.11  <---------------------- CONNECT CN_55534
2   0.11  EVENT_DISCONNECT_B ERB_Disc_B------------------------>
3   0.17  <---------------------- END_ANNOUNCEM Announcem_account


  1 Dialog's (Dlg_1) started,   0 timed out,   0 failures, t=0.00 minutes

  0 Failures, 0 Warnings
  **** E N D ****
```

4 Conclusion

Our *SSPSimulator* tool is a generic script controlled and event triggered **SINAP3** protocol generator upon TCAP level. It fulfils nearly any requirement for generating test protocol data used to drive a large IN system. *SSPSimulator* must be embedded in a special testing environment that does all preinitialization on the tested system and supplies testing data for any other external interface like so routing tree database, subscriber and provider data and any other configuration data. *SSPSimulator* fits the developer's requirements in testing for coding errors. It is also used to obtain average and peek load information as it may act as load driver for the tested system. It does not replace an overall test of the complete system in a real heterogenous network environment, as our experiences with different switching hardware show.

A Abbreviations

Abbreviation Meaning

AC	Apply Charging
ACR	Apply Charging Report
ASF	Activate Service Filtering
AT	Activity Test
CA	Call Associated
CG	Call Gap
CHP	Component Handling Primitive
CIQ	Call Information Request
CIR	Call Information Report
CON	Connect
DHP	Dialogue Handling Primitive
ERB	Event Report BCSM
FCI	Furnish Charging Information
IDP	Initial Data Packet
INAP	Intelligent Network Application Part
MTP	Message Transfer Part
NCA	Non Call Associated
RRB	Request Report BCSM
SCCP	Signalling Connection Control Part
SCI	Send Charging Information
SCP/SCF	Service Control Point/Function
SFR	Service Filtering Response
SINAP3	Siemens **INAP** standard
SSP/SSF	Service Switching Point/Function
TCAP	Transaction Capabilities Application Part

References

1. Douglas Steedman: *Abstract Syntax Notation One (ASN.1): The Tutorial and Reference*, Techology Appraisals ('90)
2. A. Krenn, P. Prohaczka, E. Galland: *Benutzerhandbuch für das Testtool SSPSIM*, Siemens Nixdorf Informationssysteme AG ('96)
3. K. Rackwitz: *SSP-/SCP-Simulator for SINAP Interface on the Siemens Protocol Tester K1197 - User Manual*, Siemens AG ('96)
4. Kirsten Sommer: *SSF-SCF IF Specification of SINAP5 V5.20 - SCF Procedures*, Siemens AG ('97)
5. Kirsten Sommer: *SINAP V5.2 Interface Specification: ASN.1 Modules and Definition of OctetStrings*, Siemens AG ('97)

Agent-Based Data Services in Future IN-Platforms

Dr. Sahin Albayrak[1,] Jens Meinköhn[2]

[1]TU Berlin, DAI-Labor, Sekr. FR 6-7, Franklinstr. 28/29
D-10587 Berlin
Sahin@cs.tu-berlin.de
[2]Deutsche Telekom AG, Technologiezentrum, Ringbahnstr. 130
D-12103 Berlin
Meinkoehn@fs.telekom.de

Abstract. The Intelligent Network (IN) architecture is an important step towards flexibly deploying and providing new telephony services. However, the IN suffers substantial shortcomings, particularly in view of the provisioning of data services and the rapid creation and deployment of new service types. This paper proposes a solution for providing data services in the IN based on agent technology. This approach also facilitates management of IN, e.g. service creation and deployment.

Introduction

The current trend towards an information society requires comprehensive multimedia communications and global information connectivity, particularly in view of the convergence of information technology and telecommunications towards telematic services. Emerging broadband network technologies over the past decade have provided the technological basis for the integration of different forms of communication. Integrated network technology and global network interconnectivity are the main driving forces behind the emergence of a global network platform as a communication facility for all types of multimedia applications. The technical solution to this has two major prerequisites, namely the development of end-systems capable of integrating data, audio and video information; and appropriate architectures for distributed multimedia applications that model and support information flows in terms of a generic platform for open distributed environments.

One of the proposed solutions for such environments is the Intelligent Network (IN) architecture which has been implemented in a number of platforms, e.g. the IN'96 platform. The IN architecture is continuously being enhanced, particularly regarding IN services as documented by the IN Capability Sets 2 and 3 which are scheduled to be finalized in 1998. However, even though existing platforms such as IN'96 offer a rich variety of sophisticated services, to date only voice-related services are available! In addition to that, the IN'96 management functionality such as service creation and provisioning is unwieldy and inefficient.

It is the goal of our work to arrive at a viable, evolutionary solution to the creation, deployment and usage of data services within the IN'96-platform by means of *agent-oriented technology* (AOT). By doing so we are providing the concepts and ideas for the foundation of future IN-architectures supported and enhanced by AOT.

In this paper we discuss our approach to extending the IN'96-platform by data services provided to the IN'96 customer and user by means of AOT. The remainder of this paper is structured as follows: chapter 2 provides a summary overview of basic AOT concepts as well as a short description of the IN'96-platform and its associated voice-based services. In chapter 3 we explain the drawbacks of IN'96 with regard to the type of services provided, and in chapter 4 our evolutionary AOT-based approach to realizing data services in the IN'96-platform is elaborated, taking into account the problem of suitable terminal equipment. The main statements made in this paper are summarized in chapter 5 where also an outlook on our future activities is provided.

Overview of Agent-oriented Technology and the IN'96-Platform

In order to provide basic background information on the subject at hand it is necessary to briefly touch the main concepts, issues and characteristics of AOT as well as the main features of the IN'96 platform. More detailed information on AOT and agent-based services can be found in e.g. [Wool-95, Ovum-94, Inamos-96a, Inamos-96b]. Here, we will only give a short description of IN'96 features; comprehensive information on the IN architecture and related standardization activities can be found e.g. in [Magedanz-96, ITU-95].

Agent-oriented Technology

The term „software agent" or, more commonly, „intelligent agent" has become increasingly popular; however, a universally accepted and generally applicable definition of an agent is not available.

Suffice it here to say that an agent is a software unit and that agent-based computation, i.e. the activities carried out by agents, can be characterized as follows:

- *intelligent behaviour.* By means of a suitable protocol, agents are able to communicate with each other. In particular, this communication capability may be used for cooperation purposes where several agents act as a team in solving a given problem. Moreover, agents may coordinate their activities, e.g. with regard to resource usage;

- *autonomous behaviour.* An agent can behave proactively, i.e. its actions are based on an internal representation of a plan and need not necessarily be triggered by an external stimulus. The agent can act in order to achieve overall goals and does not require user interactions for controlling its behaviour; and

- *decentralized control structures.* Agents can be physically as well as logically distributed and, optionally, have the capability to travel within a network according to given goals.

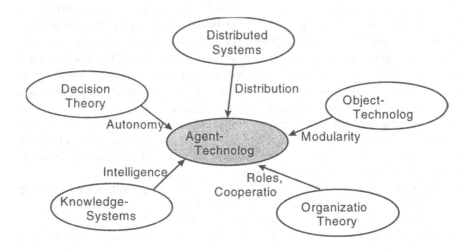

Fig. 1. Important Areas of Influence in AOT

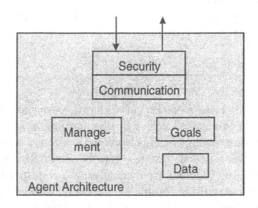

Fig. 2. Generalized Agent Architecture

AOT is a highly interdisciplinary area of interest with quite a number of different sources of input. Some of these sources and their input concepts are depicted in fig. 1. By its very nature AOT is seen as a new programming paradigm, extending object-oriented programming which evolved from structured programming which in turn superceded assembler and machine code programming.

An agent is based on a generalized architecture which encompasses major components of an agent, such as (self-)management facilities, a set of goals, a repository for data relevant to the execution and the task(s) of the agent, as well as a security and a communication component for interactions with other agents, system resources and human users (cf. Fig. 2).

Based on the generalized agent architecture, agents of different granularity can be identified. *Light-weight* agents comprise only those components and capabilities essential to basic task processing and execution. Due to its small size it is particularly well suited for tasks requiring mobility on behalf of the agent, e.g. information gathering and preliminary local filtering. Stationary *mid-weight* agents are equipped with more sophisticated capabilities and are well suited for tasks of a given but limited complexity, e.g. Personal Digital Assistants. Finally, stationary *belief-desire-intention* agents (BDI-agents) make use of e.g. sophisticated knowledge-based computation in order to provide highly complex services.

Due to their flexibility, agents can easily enhance the quality of information exchange, especially in an environment of heterogeneous and distributed communication infrastructure. Other advantageous characteristics usually found in agent-oriented approaches are the following.

- support for open system architectures by means of application programming interfaces (APIs) that are used to program agent systems;
- compliancy with established communication standards;
- provision of management and configuration functionality in dynamically changing environments;
- support for security functions like the authorization of communication parties and the encryption of transmitted data;

Therefore, the nature of agent-oriented technology blends very well with important requirements of telecommunication providers.

IN '96

The IN'96 platform offers quite a variety of features, encompassing CS-1 type services but also value-added features such as flexible tariffing of service usage and location- and time-dependent routing of calls according to customer preferences.

In particular, these value-added features can be controlled by means of personalized call management facilities for manipulating a corresponding customer profile including routing and tariffing constraints. Typically, calls can e.g. be routed to different destinations depending on the geographical location of the calling party, the date and time of day the call is made but also depending on the amount of calls currently handled at the chosen destination. Calls can be charged to the callee's account only or to both caller's and callee's account. More complex services offered by IN'96 are the VPN and UPT services enabling logically defined personalized networks and logically defined personalized communication profiles including call forwarding features and flexible, account-based tariffing.

In addition to the basic services and service features it is necessary to provide management functionality in two specific ways. Firstly, the service providers need distinguished access points to the IN platform and tools for the creation and alteration of services. And secondly, also the service subscribers have to get access to certain parameters of the subscribed services and need some way of modifying these. But again, the IN standard CS1-R is limited in a detailed explanation of such a functionality. Even though the standard encompasses the description of a 'Service Management Function' (SMF) which allows the creation and administration of services as well as the provision of statistics data, a detailed description of the particular functionality that SMF has to provide is missing. Therefore, it is up to a individual implementor of IN services to specify management functionalities.

Evaluation of Existing IN-Platforms

In this chapter we explain the need for further enhancing the currently available implementations of intelligent networks, if the obvious trend towards a universal information society is to be supported in future. Although the services described in the previous section nowadays meet the needs of the average customer, the present development of telecommunication services is limited. The following points justify this in more detail, thereby referring to the IN'96 architecture explained in the previous chapter.

However, it should be mentioned that the currently implemented standard id the revised version of the 'IN Capability Set 1' (CS1-R), which defines the currently available intelligent network services and their realization by means of service features. This standard is to be seen as only one of several steps towards a fully satisfying intelligent network architecture. The inclusion of more advanced services into it has been postponed in order to provide a publicly available reference document, such that first experiences with concrete implementations can be gained.

The previous section has described what kind of services are supported by virtue of the main concepts underlying the intelligent network architecture according to CS1-R. Here, we describe the obvious shortcomings in this.

Restriction to Voice Services

In principle, the intelligent network architecture aims at the provision of various data services, may it be voice, video or raw data streams. However, presently there are only voice services supported by the existing INimplementations. This is not surprising, because the only accepted standard for such networks is the CS1-R. Amongst other services which will be important in future but that are not yet approved, this standard does not address the transmission of data streams other than voice signals.

Of course, this must be considered a serious drawback in an information society that increasingly depends on data services. Several demanding applications rely on the transmission of more than just voice signals, as for instance all services targeting the electronic market place.

Inefficient Creation and Provisioning of Services

The services provided by intelligent network platforms are combinations of so called 'service-independent building blocks', or SIBs for short. These are aimed to provide a generic layer of functionality to be used to build specific services on top of it.

However, the standards do not attempt to describe how services are actually created, i.e. assembled of SIBs, and deployed. In the existing IN'96 architecture the creation, testing and management of services is time consuming and cumbersome. Therefore it is necessary to unify and ease these administrative tasks.

Lack of Flexibility and Means of Personalization

Again, the underlying standard CS1-R must be deemed responsible for this restriction. The means of personalization provided in its framework are limited, as can be seen by considering one particular service example, namely the 'Universal Personal Telecommunication' (UPT). Independently of the physical location or the available telecommunication environment, this service allows the subscriber to initiate and receive calls via a unique telecommunication number.

Nowadays, such a service may only be used in the physical communication environment provided by a particular intelligent network architecture. The so called 'INinterworking', which refers to the provision of transparent services across network boundaries, and which therefore requires transparency upon different providers, is not yet part of the capability set CS1-R.

It is also important that the user of such a UPT service can administrate certain service characteristics according to his personal needs. For instance, he might want to edit parameters that restrict incoming calls to a particular originator location ('originating call screening'). Currently, such a functionality can only be realized using the tone dialing facility of the peripheral in connection with voice menus that guide the user. Obviously, this kind of service actually requires the incorporation of data services to provide a convenient interface to the user.

Enhancing Existing IN-Platforms with Agent-oriented Technology

As we have seen, the IN'96-platform has a number of characteristics which obviously need to be improved in order to arrive at a platform for open distributed environments as required by the increasing customer demand for multimedia applications and services. The AOT as described above provides the basic framework for thus enhancing IN'96. In the following subsections we will explain our general control structure for distributed agent-based telematic services, as well as a 3-step migratory path towards incorporating this control structure and using it as a basis for providing data services in the IN'96-platform.

Agent-based Service Provisioning and Access

In order to enable the efficient provisioning of agent-based telematic services as well as to facilitate user access to such services, we have developed a general control structure based on a marketplace metaphor [Inamos-96a, Inamos-96b], illustrated in fig. 3, and implemented it using Java (subsequently referred to as „AOT-platform").

On the server side, a platform which is enhanced by infrastructure services and management capabilities allows service providers to install complex stationary service provider agents. By using appropriate access software (labelled „IA-Navigator" in fig. 3) on the client side, the user can create light-weight mobile agents carrying her service requests which are sent to the server platform. Note that for mobile agent migration our implementations use a TCP/IP communication protocol on top of different transport technologies, e.g. Ethernet, GSM and ISDN; also, the access software is independent of specific terminal equipment and is either installed in the terminal equipment or may be downloaded e.g. as a Java-applet. Upon arrival, the mobile user agents are authenticated and authorized by the manager agent and can then proceed to cooperate with the stationary service provider agents by means of a KQML-based cooperation protocol. The asynchronous nature of agent migration lets the user be offline while the user agents carries out its tasks; of course, conventional connection-oriented client/server communication is also possible.

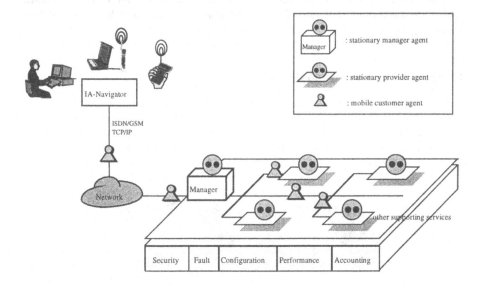

Fig. 3. Providing and Accessing Agent-based Telematic Services

AOT and, in particular, our approach to agent-based service provisioning and access offers several advantageous features:
- the possibility of accessing and using several telematic services in conjunction, i.e. the user may access several different telematic services in conjunction by means of an uniform access mechanism;
- stationary service provider agents may cooperate dynamically, depending on current user requirements, thereby enabling complex services which are not provided by any single stationary service provider agent;

- AOT supports the convergence of heterogeneous network technologies, e.g. ISDN and GSM, by using the TCP/IP protocol as a universal remote communication and transport mechanism. Thus, AOT hides the heterogeneity but lets the user take advantage of special characteristics of each network technology;

As a result of an application analysis for agent-based telematic services carried out in cooperation with T-Mobil (DTAG mobile phone operator) and OnlinePro (DTAG online service), a number of application scenarios were identified which promise a high commercial potential. Of these, the *Personal Information Manager (PIM)* seems especially interesting from an IN point of view. The PIM combines the agent-based provisioning of communication services, e.g. fax, voice and email, as well as other horizontal and/or vertical data services, e.g. remote access, file transfer, WWW, database access etc. Particularly, by means of cooperating service provider agents, the communication services may be integrated to a universal mailbox service. The PIM is freely configurable according to the user's needs and resides in the network; this results in a personalized office environment permanently accessible by mobile users independent of their current location and the currently available terminal equipment.

Other application scenarios include Electronic Commerce and Traffic Telematics which are not further elaborated here (cf. [Inamos-96a, Inamos-96b, Inamos-96c] for more details).

Using the Right Terminal Equipment

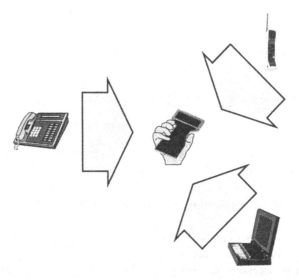

Fig. 4. Integration of Different Terminal Equipment and Functionality

Clearly, the more data intensive services are to be used in day-to-day communication, the more important becomes the problem of suitable terminals. These devices must be able to cope with large amounts of all kinds of data streams, used both for input and for output purposes. Of course, this poses a high challenge not only on the involved hardware, but also on the software that is used to present output date in a suitable manner and gather sophisticated input data from the user.

There is no doubt that the future terminal will be an integrated device capable of handling different data streams, as they are needed for voice, video and data services (cf. fig. 4). Appliances like laptops and standard telephones, or also mobile phones, are going to be combined into single, multi-purpose terminals. These will provide the functionality known from (mobile) phones as well as input and output facilities known from laptops, i.e. they will comprise a display and a keyboard.

Since there will certainly no sudden upgrade to such terminals they will rather exist side by side to traditional devices. It is therefore necessary for the service software to cope with different capabilities of different terminals. For instance, in using a data service which prompts input from an end user it will be important to consider whether input can be supplied e.g. via a keyboard or via an encoding from telephone keys.

As we have already pointed out in the previous section our basic access software for agent-based telematic services is designed to be independent of specific terminal equipment and that it is either installed in the terminal equipment or may be downloaded when needed. In fact, once the initial connection to the agent-based service platform is made e.g. by dialling a specific IN-number, tailor-made access software corresponding to the type of terminal equipment currently in use can be downloaded and used for further accessing the various agent-based services.

Agent-based Data Services in the IN'96 Platform

Our main goal is to enable the provisioning and usage of data services in an IN environment in addition to the existing voice services by employing our AOT-platform (i.e. the agent-based telematic services platform) as well as the corresponding access software. The natural residence for such a platform within the IN architecture seems to be the Service Control Point (SCP) as it is already equipped with processing capabilities.

However, in view of the fact that the introduction and implementation of an agent-based service platform within the IN'96 is a highly complex issue as well to ensure acceptance of this development with the IN'96 providers, a three-step evolutionary approach to fully integrating IN'96 and the AOT-platform is deemed appropriate. In the first step, the IN'96 serves the user only as a transport mechanism to external AOT-platforms. In order to support the user in locating desired services, in the second step the SCP is enhanced by a „yellow page"-service for intelligent service brokering but the AOT-platforms still remain external. Finally, in the third step the SCP is enhanced by AOT-platforms representing external content providers. The following subsections outline this three-step evolutionary approach in more detail.

Accessing External Data Services via IN'96

In the first step an external AOT-platform is accessed by the IN'96 user by dialling a dedicated IN number where a general prefix number represents the user request to be connected to an AOT-platform and the postfix number identifies a specific AOT-platform. Call handling is the same as with ordinary IN service calls: the Service Switching Point (SSP) recognizes the call as an AOT-platform connect request and communicates with the SCP as to which SSP and, ultimately, which AOT-platform the call should be routed to (cf. fig. 5a).

It can be easily seen that here the IN'96 serves only as a transmission medium for connecting the user to an AOT-platform and that the service logic for the agent-based telematic services is not located within the SCP, as is ordinarily the case, but within the external AOT-platform. Although this first step at least enables the provisioning of data services to IN users, unfortunately it is characterized by several drawbacks:

- *no „yellow page"-support for the IN'96 user.* The user is forced to know a priori of existing AOT-platforms and their IN numbers. As long as there are only few AOT-platforms (and related services) available, this is acceptable; but in the long run with the IN architecture foreseen as a basis for an open telematic service environment, this is not tolerable;

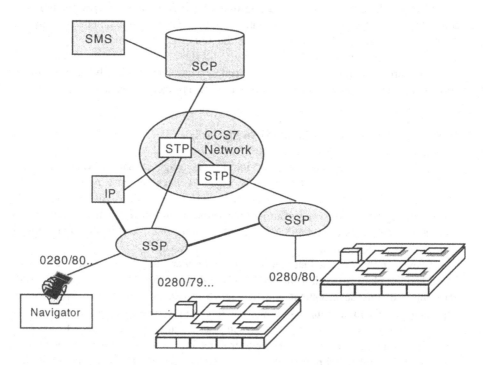

Fig. 5. Accessing External Data Services via IN'96

- *no differentiated billing by means of IN'96 billing procedures.* Of course, the call can be charged as any other IN call with regard to existing IN'96 billing procedures, e.g. duration or originating geographic region. As there is no communication between the AOT-platform and the SCP, the actual usage of individual services and service features in the AOT-platform can not be accounted for within the IN'96;

- *service management is separate from IN'96 management procedures.* Again, as with the issue of differentiated billing, no dedicated management functionality, e.g. concerning security, for the agent-based telematic services is available within the IN'96.

These drawbacks will be dealt with in the second and third evolution steps described below.

Transparent Service Access via Intelligent Service Brokering
The second step in our evolutionary approach involves the addition of service brokering functionality within the SCP. This means that the SCP call handling mechanism requires access to a database where the services and the functionality of the various AOT-platforms is represented.

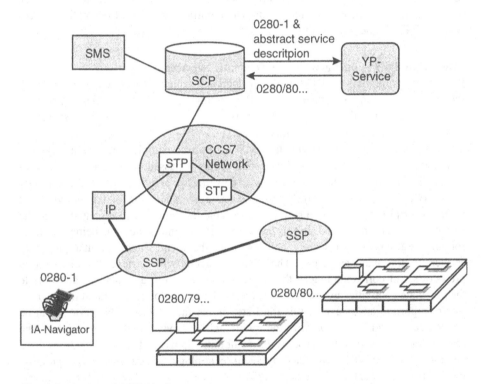

Fig. 6. Brokering of AOT-Platforms

In order to match the user's requirements with available AOT-platform functionality further interaction between the user and the IN is necessary: after the SSP has recognized the call as a request for agent-based data services via the service brokering mechanism, the SCP determines an appropriate Intelligent Peripheral (IP) to query the user for an abstract description of the specifically desired agent-based data service. The SCP returns to the SSP a routing number for that IP and instructs the SSP to establish a connection to the IP. The IP starts a dialogue with the user and prompts her to enter her abstract service description. This information is returned to the SCP via the SSP which then is able to determine a suitable AOT-platform with the appropriate service and instructs the SSP to route the call to that AOT-platform. This activity is illustrated in fig. 5b.

The benefits of this second step for providing agent-based telematic services to the IN'96 user are immediately obvious: the user no longer needs to be informed a priori about existing AOT-platforms and the agent-based services provided by them, but instructs the IN'96 on the type of service desired. This substantially enhances the attractiveness of the IN'96 and enables additional flexibility to the IN'96 provider with regard to individual specific AOT-platforms connected to the IN.

Unfortunately, differentiated billing with regard to the agent-based services actually accessed by the user is still not available. Management of the agent-based services in the AOT-platforms is still outside the IN'96. This situation will be remedied by the final evolutionary step of integrating agent-based telematic services with the IN'96 as outlined below.

Integrating IN'96 and Agent-based Telematic Services
For finally integrating the IN'96 with our approach to agent-based telematic services, the SCP needs to be enhanced in the following way: the AOT-platform is incorporated within the SCP such that individual service features as represented by SIBs are modelled as feature provider agents.

An IN service is created via the SMS by using a simple formal language for formulating the desired service's functionality; this is passed to the manager agent of the AOT-platform which then initiates a cooperative activity among the provider agents (which represent individual service features). The result of this activity is a protocol script for cooperation stored in a suitable database and instantiated when the service is actually requested. In this case, an SSP recognizes a call as being a request for an agent-based data service and forwards the identification of that particular service to the manager agent. The manager agent will retrieve the cooperation protocol script corresponding to the requested service from the database and initiate the associated activities by the provider agents. This situation is illustrated in fig. 5c.

An important advantage of this third step is the fact that new IN services can be created easier and faster than with the traditional IN service creation procedures. The reason for this is that the process of combining individual service features to a complex new service is automated by agent technology, i.e. through the cooperation mechanisms inherent in the feature provider agents of the AOT-platform and the resulting cooperation protocol script. This is opposed to the traditional way of creating IN services where SIBs representing service features must be combined manually by using a graph-based creation tool. This is an extremely error-prone activity which can be substantially simplified by using the automated creation facilities of the AOT-platform.

Furthermore, because of the fact that the management functionality for agent-based telematic services is now within the SCP, the differentiated management information on service usage, e.g. billing information, can be collected and further processed by the SCP. Also, management functionality other than just accounting and billing for the agent-based telematic services is now available within the SCP.

Note that additional AOT-platforms can be connected directly to the main AOT-platform by installing there a dedicated agent which serves as a direct gateway to that external AOT-platform. The dedicated agent collects relevant management information from the external AOT-platform and provides it to the SCP for further processing. This is a straightforward mechanism for easily scaling the number of agent-based telematic services available to the IN'96 user.

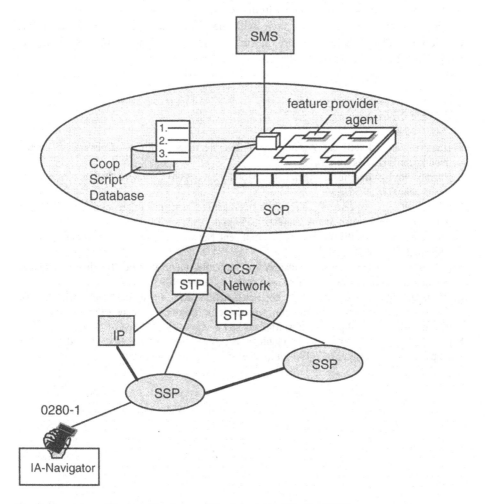

Fig. 7. Agent-based Service Creation and Provisioning within the IN'96

Concerning the ubiquitous and still unsolved problem of undesired service feature interactions, this approach enables the detailed examination of possible solutions: for example, individual feature provider agents could cooperate at run-time in order to identify and resolve undesired service feature interactions.

Conclusion

In this paper we have introduced the concepts of AOT and its main advantages regarding telematic services and we pointed out some major drawbacks of existing IN platforms such as IN'96. By using our approach to agent-based service provisioning and access, we have explained a viable means of integrating data services with IN platforms in a three-step evolutionary way. Additionally, we have shown how the management of such data services including the improved creation and provisioning of IN services can be incorporated within the IN.

Through our approach we have laid the foundations for a possible future development of the IN architecture, particularly in view of the upcoming upgrade of the IN'96 platform to the IN'98 platform.

References

1- [Inamos-96a]   Intelligent Agents for Mobile Services-I", Project Deliverable, contact meinkoehn@fz.telekom.de
2. [Inamos-96b]   Intelligent Agents for Mobile Services-II", Project Deliverable, contact meinkoehn@fz.telekom.de
3. [Inamos-96c]   Anwendungsanalyse für Intelligente Agenten in Telematikdiensten", Project Internal Report, contact meinkoehn@fz.telekom.de
4. [Inavas-96a]   Intelligent Agents in Intelligent Networks-I", Project Deliverable, contact meinkoehn@fz.telekom.de
5. [ITU-95]   ITU Recommendation Q.12xx series, General Series Intelligent Networks Recommendations Structure" (Revised Version), Geneva, 1995
6. [Magedanz-96] Magedanz, T., Popescu-Zeletin, R., Intelligent Networks: Basic Technology, Standards and Evolution", International Thomson Computer Press, 1996
7. [Ovum-94]   Intelligent Agents: Markets and Technology", Ovum Ltd., London, 1994
8. [Wool-95]   Wooldridge, M., Jennings, N., „Intelligent Agents: Theory and Practice", http://www.doc.mmu.ac.uk/STAFF/mike/ker95/ker95-html.html, 1995

Visualisation of Executable Formal Specifications for User Validation

M. B. Özcan, P. W. Parry, I. C. Morrey and J. I. Siddiqi

Sheffield Hallam University,
Computing and Management Sciences,
Computing Research Centre,
City Campus, Sheffield S1, 1WB, UK.
Email: M.B.Ozcan@shu.ac.uk

Abstract. This paper reports on research work to facilitate the user validation process in an application-orientated fashion based on executable formal specifications. It is part of an ongoing effort to move towards quality requirements via graphical visualisations of formal specifications. It builds upon previous work that supports the animation of Z specifications in a LISP-based environment called *ZAL* (Z Animation in LISP). In addition, it embodies a visualisation system called *ViZ* (Visualisation in Z) which enables the comprehension, clarification and validation of executable formal specification notations. Technology provided by ViZ allows software developers to choose an appropriate representation of objects used in an executable formal specification and create dynamic and/or static animations of these objects in an interactive and iterative fashion. ViZ provides a generic visualisation model to capture the process of visualising static and dynamic behaviour of a ZAL specification. This paper outlines our approach, details ViZ and illustrates its application in a real-world setting.

1 Introduction

In this paper, we argue that an eclectic approach that offers an effective combination of formalism and pragmatism may encourage software developers to move towards software engineering practices necessary for software systems that satisfy users requirements. Many have advocated the use of executable formal specifications for the construction of prototypes to validate software requirements with the users at an early stage through feedback [1]. An executable formal specification as a prototyping technique bridges the gap between traditional software prototyping and formal methods, which have quite different origins, and until recently were seen as disparate developments. Equally important, it maximises the strength of formal specifications and prototyping, and minimises their weaknesses. Furthermore, it facilitates the systematic development of a clear, concise, precise and unambiguous specification of a software system which is characteristic of a formal approach, but absent in traditional prototyping. It also enables the developer to execute the formal specification at an early stage, which is absent and often strongly prohibited in formal methods.

Despite their benefits, executable specifications are not without their problems. In our opinion, the key problem, which is the focus of the remainder of this paper, is that they are often ineffective in the user validation process. Executable formal specifications have traditionally been used as an effective tool to facilitate developer validation; that is, a developer can, via specification execution, either individually or in a peer review format, explore the consequences of the specification. Hence, its use in requirements validation is often not user orientated, which is likely to reduce the effectiveness of this eclectic approach. This is exacerbated by the fact that the execution behaviour of a prototype based on formal notations may not always be comprehensible to users. This may in turn render the requirements validation process less effective.

This paper reports on research work to facilitate the user validation process in an application-orientated fashion based on executable formal specifications. It is part of an ongoing effort to move towards quality requirements via graphical visualisation of formal specifications. It builds upon previous work that supports the animation of Z specifications in a LISP-based environment called *ZAL* (Z Animation in LISP) [7]. In addition, it embodies a visualisation system called *ViZ* (Visualisation in Z) which enables the comprehension, clarification and validation of executable formal specifications. Consequently, our work can be regarded as a quality assurance process in which the initial requirements are formalised, and interactively demonstrated and clarified to produce an improved set of quality requirements. The remainder of this paper details our approach and illustrates its application in a real world setting.

2 Background Work

Our initial work focused upon the development of a software system for specification, construction and animation in Z that can be incorporated into any software development model that has early activities corresponding to requirements capture and formalisation [7]. The system enables software developers to formalise and construct requirements at the most abstract level and assist them in clarifying existing requirements as well as discovering "unknown" ones. The system's main objective is to facilitate the developer validation process by animating Z specifications in an executable language (in our case LISP).

The approach, supported by tools, can be used to receive early feedback in a timely and low risk fashion. The toolset has essentially two components: *TranZit* is a full-screen editor, syntax and type checker for constructing Z specifications. It also incorporates a transformation engine that generates the corresponding executable representation of a given Z specification for animation in an extended LISP format. *ZAL* is a LISP-based animation environment which provides the execution mechanism for animation. It provides extensions to LISP to form an animation environment. A close correspondence between the Z notation and the ZAL notation is preserved. This enables abstractions achieved in Z to be retained, which in turn makes the translation process relatively mechanical. The hallmark of ZAL is to view it as a generic animator which models Z constructs rather than any particular specification; the subset of Z that can be animated is that for which equivalent constructs have been developed in ZAL.

The process for the validation of requirements using this toolset is carried out in four stages (Fig. 1). First, a formal specification is defined using TranZit after the initial requirements have been elicited. Second, this specification is forwarded to the transformation engine in order to build an animation, that is the specification is converted by the transformation engine into an executable representation suitable as an animation for developers. Third, the LISP-based executable version is forwarded to the ZAL system to execute the animation to allow developers to interact with it in order to demonstrate various properties of the original specification. Finally, once the developers have validated the specification with respect to their understanding of what is required, the execution behaviour of the system is shown to users for early feedback. As the users work with the prototype system, they might report problems, detect opportunities for new features, and might make enhancement requests for these new features. It is intended that each evolution of the specification that is synthesised by the requirements engineer should be formally recorded using Z, and that the specification should then be animated using ZAL and executed with the other requirements engineers participating in the validation process. The specification can then be reasoned with and expected behaviour confirmed.

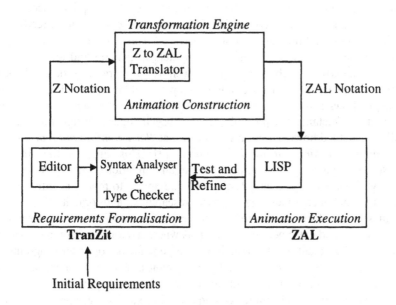

Fig. 1. Validation of Requirements in the ZAL System

The process, along with the toolset, assists developers in gaining greater confidence in the constructed specification. Within the ZAL environment it is possible for a developer, or more likely more than one developer in a peer review manner, to improve their understanding of both the specification and the requirements by animating various specification scenarios. Although our approach is not merely limited to developer validation, our experience suggests that its use for user validation is less effective. The dialogue that takes place between the system and its users does not always effectively reflect the external behaviour of a software system from users' perspective. This is largely due to the fact that the animation does not provide a full transparency of the underlying specification notations (such as set notations) to the users. Consequently, a fundamental decision has been made to establish the requirements of an interface to the animation system, and the use of visualisation techniques to promote an effective medium through which the requirements of a system can be negotiated. As a result, an application-oriented process has been developed, together with a visualisation system to act as a catalyst to the process and to provide support for the dynamic graphical animation of ZAL notations. The remainder of this paper focuses upon such new features of the ZAL system.

3   Visualisation and Its Application to Formal Specifications: Relevant Issues

The use of graphics and animation techniques have been applied successfully to virtually all major areas of computing from word-processing and spreadsheets to operating systems and human-computer interfaces. This can be attributed to a rapid increase in the graphics capabilities of computers in general and an equally rapid decrease in the cost of such technology, as well as the increase in demand for such features from all aspects of the user and development communities. The use of visual technology allows complex concepts and information to be presented in ways which are easier to understand. The physiological and psychological reasons for this are not yet fully understood however, but it is evident that "*human information processing is clearly optimised for pictorial information*" [3]. This suggests that in certain circumstances a picture (or set of pictures, i.e. an animation) can be a more suitable medium with which to present information. The ability to represent information by means of visual techniques has allowed researchers and practitioners in other disciplines to transform data pertinent to their fields into pictorial representations. This has been found extremely useful in areas where quantities of complex data have resulted from the execution of simulations, experiments, or computer programs, such as meteorology, medical science, and aerodynamics. This in turn allows patterns inherent in the data or other useful observations, to be made apparent.

We argue that visualisation techniques can equally be applied to formal specifications with equal efficiency. The immediate benefit can be the facilitation of the communication between the developers and the customers. In theory, this can be achieved by communicating complex information and concepts effectively via the use of powerful images and animation techniques instead of pure textual notations. In practice, however, visualisation techniques should be regarded as a complementary approach, which strengthens the use of formal methods, rather than a complete alternative. Application of visualisation to formal specifications requires the use of a process which maps abstract representations to concrete representations which effectively captures the external execution behaviour of the abstract representations [4]. Unlike a programming language translator, the translation between the two representations is bi-directional so that users can directly interact with a software system. There are a number of issues that should be considered to support this process. In this paper, we will address the following three issues:

Type of Representation.   The type of representation employed in the visualisation process may have two attributes. These are direct and abstract respectively. Direct representations provide a realistic representation of an object or system under consideration which is close to its real-world counterpart, such as a video-clip or a photographic image. This type of representation is best used when trying to communicate highly detailed information where a diagram could possibly hide some of the content. On the other hand, abstract representations provide an abstract image of the underlying information, and thus providing an appropriate mechanism to emphasise certain details while hiding others. Examples of abstract representations, are charts or graphs, or diagrams which show relationships between information such as trees or networks.

In order to represent a formal specification notation such as a ZAL schema, the system state, its invariants, input and output variables and local variables need to be mapped onto real-world objects and values to animate this schema. For instance, ZAL schemas used to specify an air-traffic control system may be represented in terms of objects such as planes, a set of runways and an air-traffic control tower, etc. The type of representations (i.e. direct or abstract) suitable for these objects is ultimately a choice for developers. However, in practice, it could be dependent upon the nature of a problem domain and its complexity, the preferences of users and the degree of their expertise in the problem domain.

Moreover, under certain circumstances in order to enrich the overall visualisation of a formal specification, it may be necessary to augment the visualisation with related images. The notion of relatedness corresponds to 'context' which can be thought of as the 'scene' or 'situation' in which a formal specification is interpreted. If the representation of a *context* can be closely coupled with the overall behaviour of a formal specification under consideration, the visualisation of the formal specification with its augmented form may facilitate the users' comprehension of this behaviour (e.g. an image of a library at the background may augment the visualisation of a formal specification that records the subscription of members to the library). In order to develop contextual representations, it is important to understand the essential characteristics of the formal specification such as the overall system's requirements and the composition of the tasks which make up these requirements.

Nature of Animation.   This issue is concerned with capturing the process of visualising dynamic behaviour and may be addressed in terms of a generic visualisation model. The level of animation has two attributes. These are *static* and *dynamic* respectively. *Static* refers to an unchanging, still image, such as the diagrammatic notations data flow diagrams. This form of representation is useful when attempting to convey relationships between objects, and for analysing the structural properties of systems or data. *Dynamic representations*, however, imply a continuously changing set of images, which correspond to some execution process undergoing successive changes. These types of representation are useful when depicting, analysing, and understanding dynamic processes [2]. It is widely accepted that effective validation is performed when the users observe a dynamic representation of the system's requirements [8]. In this context, the notion of "animation" (in the visual sense) becomes a prominent issue. Aspects such as what to visualise (e.g. should a formal specification as a whole be visualised or is it sufficient to visualise its components individually?), and how to visualise it, are important aspects of dynamic representations and should be considered.

Semantic connection between a formal specification and its visualisation. It is imperative that precision introduced by the formal methods approach is carried over to the prototyping and validation stages through visualisation, in the sense that the representation of a particular software requirement based on a visual technology should not deviate from the actual meaning of the requirement represented by a formal specification. Unlike the formal methods approach, visualisation is a subjective activity which largely depends on the visualisers' creativity and imagination as well as their interpretation of what a formal specification is for and how to convey the formal meaning through animation. It is, therefore, necessary to scrutinise the relationship between a formal specification and its corresponding visualisation. From the formal specification point of view, the formalisation of software requirements should be carried out completely independently of any particular implementation concern including its visualisation. However, the process of visualising a formal specification should not be carried out independently of the specification to avoid the fabrication of ad-hoc visualisations that are not related to the formal specification. Instead, it should be done systematically and be driven and dictated by the underlying formal notation for which the visualisation is used.

4 Process and Toolset

Our starting point is a prototype system constructed based upon a ZAL animation of a given Z specification as shown in Fig. 2. In order to present the external behaviour of the prototype system in an application-oriented fashion, a scenario-based approach is adopted based upon dynamic graphical visualisations. This choice is appropriate for our purposes since scenarios are instrumental to describing and clarifying the relevant properties of an application domain, uncovering 'unknown' requirements, and challenging assumptions made about the system's properties and constraints [5]. Furthermore, scenarios provide a framework for investigating a large specification in stages, i.e. in terms of a series of related scenarios, each of which corresponds to a particular fragment of the specification.

Scenarios are first identified directly from the formal specification. These may initially be derived from the state model and invariant of the system which can subsequently be expanded to include individual or a combination of schemas in the

ZAL specification which are of interest to users as well as developers. This involves the identification of *objects* that play a role within a scenario and their *actions* in the context of the scenario.

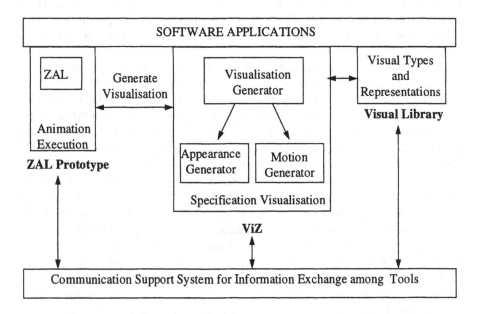

Fig. 2. Visualisation of ZAL Specifications

Our visualisation technology provided by ViZ allows software developers (or *visualisers*) to choose an appropriate representation of objects and create dynamic and/or static animations of these objects in an interactive and iterative fashion. ViZ provides a generic visualisation model to capture the process of visualising static and dynamic behaviour of a ZAL specification. It is based upon the notion of a state and is described in terms of visualising the present state of a system before the execution of a ZAL specification and the modified state of the system after the execution of the specification. In this context, visualisation of the formal specification is a composition of *visual representations* (i.e. direct or abstract) and the corresponding *motions* (i.e. static or dynamic). Consequently, a complete visualisation of a ZAL specification involves the fabrication of a composition containing all the *representations* and the corresponding *motions* for the states of the system. This is carried out as follows:

Graphical *representations* of the main objects used in the ZAL specification (such as the system state, input and output parameters of a schema as well local variables) are produced using an Appearance Generator. This is a general purpose software tool which provides a systematic framework for incorporating a set of commercially available off-the-shelf software tools that are used to generate a wide range of representations including conventional text, icons, audio and visual representations (such as sound, video animation and photographic quality pictures) and bitmap images. The role of the Appearance Generator within ViZ is therefore to provide an abstraction mechanism over the nature of these tools so that techniques required to construct a particular representation of a ZAL specification is not constrained by the specific nature of software tools. For instance, any graphical drawing tool can be used to create icons and photographic quality pictures providing that it creates the appearances in a format that can be interpreted by ViZ (e.g. a bitmap).

Static and dynamic behaviour of the main objects are produced in terms of *motions* using a Motion Generator. An analyses of what a particular object does in a particular application domain may indicate the form of static and dynamic animation needed to model the corresponding behaviour. For instance, in the context of the air-traffic control system, a plane is expected to take-off and land on using one of the available runways while communicating with the air-traffic control. The Motion Generator in this context allows a dynamic on-screen route on which a particular object will travel to be defined and generated. The tool is graphically-based and allows a developer to 'draw' a required on-screen route interactively on the screen. If the behaviour of an object has a static nature, the Motion Generator simply allows the static behaviour of the ZAL specification to be displayed on the screen (i.e. in terms of the state of the system affected by the execution of the ZAL specification).

Moreover, a Visualisation Generator is used to fabricate a complete visualisation of the specification by composing together the visualisation of the states of the specification as described earlier in terms of *representations* and *motions*. This visualisation is then related to the specification concerned so that during its execution by the ZAL system, ViZ will be able to 'play' its corresponding visual animated representation for user validation. To achieve this, the Visualisation Generator produces a unique identifier for each visualisation generated. This can then be assigned to the ZAL specification with which it is associated. Thus, each mathematical expression within the ZAL language is augmented with the notion of a visual identifier to facilitate its visualisation. Note that incorporating a visual reference to the specification does not necessarily bias it towards a particular implementation concern since the visualisation process takes place after the specification has been written. Furthermore, the Visualisation Generator is used as a 'semantic checker' between the type of a ZAL expression/object representation and the type of the corresponding visual representation. In other words, it does not allow an arbitrary representation to be assigned to a specification. Each visual representation created by the Appearance Generator is regarded as an instantiation of a type which is defined and specified in ZAL. For instance, different images of a book, such as a bit-map image or a video-clip, are simply different visual values of the underlying type book. In this context, the Visualisation Generator simply ensures that the visualisation of a book in a ZAL specification can only be obtained from the visual images associated with this type defined in the specification. The visual values and the type information that represent them are stored in a Visual library as shown in Fig. 2. Note that the organisation of the Visual library and the details of the visual type domain are beyond the scope of this paper and thus will not be elaborated here.

5   An Illustrative Case Study

To illustrate our approach it is applied to a real-time safety-critical application- a Water Level Monitoring System (WLMS) [6]. The informal requirements of this system are as follows:

> *This specification is concerned with the operation of a Water Level Monitoring system which might be used in a safety-critical system involved in steam generation, for example in a power plant. The system consists of two reservoirs; one serving as a steam generation vessel, and the other as a source of water. Under normal operation, water is pumped from the source into the steam generating vessel where it is evaporated. The pump transferring water to the generating vessel and the pump controlling the rate of steam generation in regulated by a control system termed the WLMS.*
>
> *The WLMS monitors and displays the level of water in the stream generating vessel. When the water level is too high or low, the WLMS issues visible and audible alarms and shuts down the pumps. Pumps are also shut down if the WLMS itself fails either due to external faults (such as failure of the water level detector) or internal faults in the WLMS computer. Internal faults are detected by an external watchdog which receives a periodic KICK from the WLMS. If and external faults is detected by the WLMS or the watchdog fires, the WLMS shuts the system down by turning off power to both pumps.*

In this section, only a minimal commentary on the specification will be presented for illustration purposes. Consider the specification of the top level component, pump environment. This component controls the pump switch which is responsible for controlling the power to the pumps. This is determined by the current state the operating and failure modes, the buttons, and the state of the power supply. The following events dictate the closing (pump on) and opening (pump off) of the pump switch: If the power is off then the pump switch is open (the pumps are off). If the operating mode is *Operating* and the failure mode is *AllOk*, and the reset button is released (i.e. it has already been pressed) then the pump switch is closed. If the system enters any mode other then *Hardfail, Operating*, or *Shutdown*, then the pump switch is open. The formal specification of this component in Z is given as:

WATCHDOGTYPE : : = uninit | operate | shut

ONOFFTYPE : : = on | off

PUMPSWITCHTYPE : : = open | closed

SHUTDOWNSIGNALTYPE : : = go | stop

ALARMTYPE : : = silent | audible

```
┌PumpEnvironment────────────────────────────────────
│ powerNow? : ONOFFTYPE
│ ControlSignals!
│ StoredVar'
│ ─────────────────────────────────────────────
│ (powerNow? = on ∧ shutdownSignal' = go ∧ watchdog! = operate ⇒
│    pumpSwitch! = closed)
│ (powerNow? = off ∧ shutdownSignal' = stop ∧ watchdog! = shut ⇒
│    pumpSwitch! = open)
│
└────────────────────────────────────────────────
```

The Pump Environment schema can now be translated into the corresponding ZAL form using the tools described in Sect. 2. This is shown as:

```
(make powerNow? 'on)
(make shutdownSignal 'go)
(make watchdog! 'operate)

(schema PumpEnvironment
   :! (pumpSwitch!)
   :show (pumpSwitch!)
   :predicate
   (and
       (if (and
               (eq powerNow? 'on)
               (eq shutdownSignal 'go)
               (eq watchdog! 'operate)
           )
           (make pumpswitch! 'closed)
           t
       )
       (if (and
               (eq powerNow? 'off)
               (eq shutdownSignal 'stop)
               (eq watchdog! 'shut)
           )
           (make pumpswitch! 'open)
           t
))))
```

The above specification defines a scenario where the power is 'on', the shut down signal is set to 'go', and the watchdog is set to 'operate'. The system state is initialised in such a way as to reflect these conditions. In accordance with the WLMS's formal specification given above result should be that the pumps are turned on (i.e. pumpswitch! is set to 'closed'). Executing this specification using a command line interface produces the following result:

```
>(*P (PUMPSWITCH! CLOSED))
```

It may be perfectly feasible for software developers to use a command line interface to validate the WLMS specification in a peer review format. However, this is inappropriate for user validation since the animation does not provide a full transparency of the underlying specification notation. Unlike this approach, the specification visualisation provided by the ViZ system can improve the comprehension of the behaviour of the ZAL specification under investigation. Fig. 3 illustrates a visualisation of the above scenario as a result of the above input conditions as an alternative to the command-line-based animation.

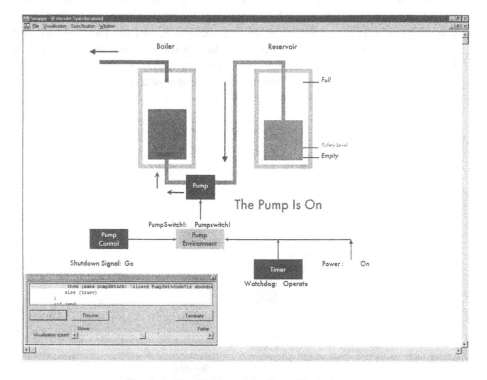

Fig. 3. A Visualisation of the Pump Environment

Using this animation, the users are shown that the system is expected to close the pump switch. The visualisation shown provides a one-for-one correspondence between the ZAL specification and the corresponding visualisation. For instance, the input parameters, powerNow?, shutdownSignal, watchdog! in the ZAL specification are represented as static objects in the visualisation with their initial values which directly come from the ZAL specification. In addition, the visualisation is augmented with a contextual representation to facilitate the users' comprehension of the consequence of closing the pump's switch. The visualisation of the Reservoir, Boiler and Pump, together with their connection correspond to such a context. An internationally recognised colour scheme can also be employed to illustrate the water temperature in different containers to enhance the users' comprehension further. This visualisation can be used to promote users' involvement and can effectively be used to acquire further scenarios from the users' to validate the specification under investigation. Therefore, the visualisation becomes a catalyst to the user validation process by assisting the identification of scenarios rather than just a front-end process support mechanism. For instance, the users may suggest to change the available parameters of the Pump Environment to observe the corresponding behaviour.

Consider a scenario where the power is 'off', the shut down signal is set to 'stop', and the watchdog is set to 'Shut'. Changes in input values are first carried out at the specification level by initialising the system state to reflect these conditions. The expected behaviour of this scenario is that the pump should stop operating. This behaviour is confirmed with the visualisation shown in Fig. 4. Note that the contextual representation reinforces this behaviour by illustrating the stoppage of the water flow upon opening of the pump's switch.

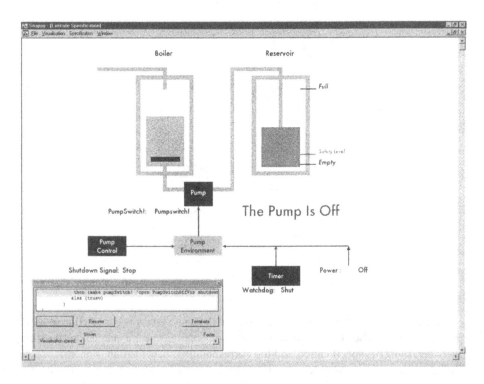

Fig. 4. A Visualisation of 'Pump is Off'

Moreover, the visual prototypes as depicted in Fig. 3. and Fig. 4. can play an important role in identifying deficiencies in the specification. Consider a situation where the users want to change only one of the three available input parameters to the Pump Environment such as Poweron? The natural expected outcome of this scenario is that the pump's switch should be opened due to power failure. However, the execution of this scenario reveals that the pump environment does not in fact open the pump's switch as shown in Fig. 5.

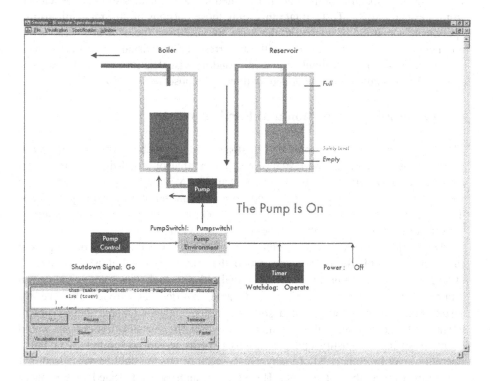

Fig. 5. Impact of 'Power failure' on the Pump Environment

When carefully examined, the formal specification reveals a mistake:

```
(if  (and
            (eq powerNow? 'off)
            (eq shutdownSignal 'stop)
            (eq watchdog! 'shut)
      )
      (make pumpswitch! 'open)
      t)
```

It suggests that it is necessary to satisfy all the input conditions to open the pump's switch. In reality, however, it should be sufficient to satisfy any of the conditions as indicated in the informal requirements (i.e. turning the power off on its own independently of the other two input parameters should be sufficient to open the pump's switch!). Hence, the specifications should be corrected by substituting 'and' with 'or'.

The same approach can equally be applied to validate the remaining schemas. Large specifications can be validated in stages in terms of a series of scenarios. For instance, the users can be demonstrated that if the water level drops below the safety level then the alarm should be audible. This visualisation should enable the users to raise a number of issues associated with the consequences of raising the alarm. Allowing large specifications to be investigated in stages enables software developers to master complexity. To facilitate this, only those portions of an executable formal specification that are of interest to the users are visualised (i.e. it is not necessary to augment each ZAL expression with a corresponding visual identifier). Those specification portions without any corresponding visualisations are still executed though their execution behaviour cannot directly be observed via ViZ.

6   Concluding Discussion and Further Work

This paper described an approach to visualisation of executable formal specifications based on the Z notation. The visual prototype was that of a real-time safety-critical application- a Water Level Monitoring System (WLMS) and was developed from a formal specification written in Z derived from an initial set of informal requirements. This section reflects upon our conclusions and addresses several key issues including the effectiveness of the visualisations of the WLMS system.

The WLMS system provided a rich foundation with which to apply our approach including the background work on which the ViZ approach builds. The WLMS system lends itself naturally to being visualised due to its inherent event-driven nature. It is these events and their dynamic nature that provide the cue for visualisation. In other words, if a software system did not possess these types of events and readily visualisable dynamic nature then the visualisation process would be more difficult to justify and perhaps its benefits would be limited. A complex operation may be comprised of many events, each of which could be visualised, but not necessarily at the same time. The scenario approach advocated in the paper allows individual fragments of a specification to be investigated and visualised in turn. The first research issue in this context is addressing the problem of 'composition' of individual fragments to demonstrate the overall behaviour. The following concerns should further be investigated to address this issue: the structure of the whole operation, how each sub-operation relates to the whole and how these sub-operations combine to produce the behaviour required by the whole operation.

Another aspect which deems this type of system as being amenable to visualisation is that each event should map onto a corresponding real-world event. Real-world events are less likely to be based upon abstract processing algorithms and are therefore easier to visualise. For example, compare the act of turning the pump's switch on with that of calculating the square root of a given argument. The former is easily recognisable and maps onto a definite real-world event, while the latter is more abstract. Note that the actual complexity of the operation is irrelevant, but it is the level of abstractness which dictates the potential for visualisation. In this context, it can be argued that the visualisations used in the WLMS system comprise of both real-world and abstract forms. The more abstract visualisations attempt to depict events such as the timer's watchdog send the 'operate' signal (refer to Fig. 3). It may be argued that visualising the contents of system variables such as the watchdog before and after the event takes place, the event and its effects can be sufficiently presented to the viewer. It is essential therefore that the representations used attempt to present the function and contents of variables in an easy to understand form. Using consistent representations for the various objects, such as pump, pump control, etc., helps this. However, some may argue that these visualisations are too abstract, and still rely upon the viewer being able to comprehend the representations and meanings given for the various objects associated with the WLMS system. Therefore, a more concrete approach might be more effective. Such an approach can be based upon sequences of video images of these objects. The technology provided by the ViZ environment enables such images to be incorporated in visualisations easily.

Another important issue is the role of contextual visualisations. Our experience suggests that such visualisations have an important role to play in facilitating the users' comprehension of a formal specification. In the context of the pump environment, the visualisation of contextual objects such as the reservoir and the boiler simply reinforce the consequence of turning the pump's switch. However, we also found out that the contextual visualisations can easily distract the users from the main points of a visualisation. This is shown in Fig. 6.

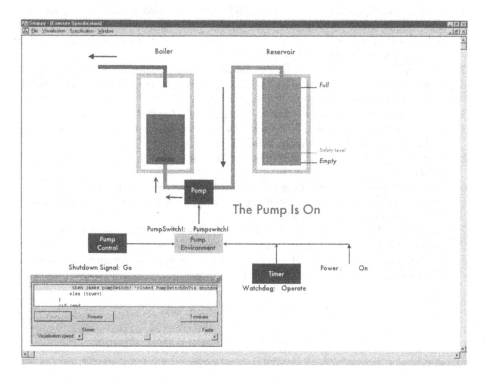

Fig. 6. Negative Impact of Using Contextual Representations

When this visualisation was shown to the users they began to ask questions about the behaviour of the contextual objects instead of asking questions about the real objects under investigation (e.g. Pump control, timer, etc.). One of the questions that was raised during the prototype evaluation stage was that 'What if the water level exceeds the maximum safety level point?' The natural answer to this question would be that the alarm would be audible so that the WLMS would shut down the pumps. However, this behaviour cannot be demonstrated in this particular visualisation since the pump environment under investigation does not check the water level (i.e. the water level is not an input to the pump environment).

We are currently investigating the role of contextual objects and their impact on the users' comprehension and under what circumstances the contextual visualisations should or should not be used.

Acknowledgements

The authors acknowledge Richard Hibberd and Graham Buckberry who developed the ZAL system. The authors also acknowledge David Cooper who wrote the Z specification on which the ZAL specifications in the case study are based.

References

1. N. E. Fuchs, 'Specifications are (preferably) executable', *Software Engineering Journal*, 323-334, September, (1992).
2. Jeff Kramer, Keng Ng , 'Animation of Requirements Specifications', Software- Practice and Experience, 18, (8), 749-774, (1988).
3. B.A. Myers, R. Chandhok, 'A. Sareen Automatic Data Visualisation For Novice Pascal Programmers', IEEE Workshop on Visual Languages, p192-198, IEEE Computer Society Press, October 10-12, Pittsburgh, PA, USA 1988.
4. P. W. Parry, M. B. Ozcan and J. Siddiqi, "The Application of Visualisation to Requirements Engineering" , Proc.8th International Conference on Software Engineering and Its Applications, Paris, France, 699-710, (1995).
5. C.Potts, et. al., 'Inquiry-based Requirements Analyses', *IEEE Software*, 11, (2), 21-32, (1994).
6. A. J. van Schouwen, 'The -7 Requirements Model: Re-examination for Real Time Systems and an Application to Monitoring Systems', Technical Report 90-276, Queens University, Kingston, Ontario K7L 3N6, (1991).
7. J. Siddiqi, I. Morrey, C. Roast and M. B. Ozcan, 'Towards Quality Requirements via Animated Formal Specifications', *Annals of Software Engineering*, 3, 131-155, (1997).
8. A. Tsalgatidou, 'Modelling and Animating Information Systems Dynamics', *Information Processing Letters*, No 36, 123-127, (1990).

Design and Evaluation of a Visual Formalism for Real Time Logics

M.Lusini[1] and E.Vicario[1]

Dipartimento Sistemi e Informatica, Università di Firenze
e-mail: {lusini,vicario}@aguirre.ing.unifi.it

Abstract. A visual formalism for the presentation of a real time logic is introduced, motivated, and evaluated. The visual formalism has been designed following a user-centered usability engineering process, targeted to the students of higher education courses in software engineering. On the one hand, heuristic design was applied to maximize consistency, i.e. to minimize the complexity of the visual metaphor mapping textual sentences to the visual representation. On the other hand, individual metaphoric assumptions were defined by prototyping and exposing alternative graphical representations to a representative sample of the target community of expected users.

The resulting notation has been implemented within a syntax-directed interactive editor which integrates the visual presentation with the conventional textual notation. The editor has been used to carry out a competitive user-based evaluation of the usability of textual and visual representations, by carrying out a readability test on a larger sample of representative end-users.

1 Introduction

Time affects the development of a large class of reactive systems, whether because of explicit real time constraints [8] [23], or due to sequencing limitations which implicitly result from timed behavior [10]. Engineering the verification of both the factors is largely motivated by their inherent complexity, often exacerbated by the criticality of the application context [25] [22]. To support systematic approaches to this verification, a number of real time logics have been proposed, which extend well consolidated untimed temporal logics [21] [16] [7] to encompass expression and checking of both qualitative sequencing limitations and quantitative timing constraints [3] [2].

Up to now, design and evaluation of real time logics has been mainly focused on the critical contrast of expressivity versus verifiability, i.e. the capability to express relevant facts versus the possibility to automatically check the satisfaction of expressions [4]. While this has permitted to develop expressive power and checking algorithms up to a substantial maturity, it has also led to a major increase of notational complexity, which now comprises a major hurdle for the actual industrial usage. To get over this hurdle, it is now necessary that both expressivity and verifiability find a suitable trade-off with the third emerging need of *usability*. Visual formalisms may largely help in finding one such trade-off by supporting effective presentation of complex notations, so as to facilitate syntactic parsing and to establish interpretation metaphors allowing intuitive understanding of expression semantics [11] [12].

The use of a visual notation for the expression of temporal logic sentences was first proposed in the definition of the untimed *Graphical Interval Logic* (GIL) [15] [13]. By relying on the basic temporal construct of a time interval delimited by two state transitions [1], GIL can build on the familiar representation of *Timing Diagrams* to support intuitive understanding of time progressions and partial orders among time intervals. Recently, the representation of GIL has been extended into RT-GIL to encompass expression of real time constraints [17]. Unfortunately, the visualization approach of GIL cannot be extended to the large variety of temporal logics which do not explicitly include the interval construct. This type of logics, which attracts most of the research and experimentation effort in the community of computer aided verification, usually rely on the lower level temporal construct *until*. Visual formalisms for temporal logics based on this construct have been proposed in [9] and [14]. In [9], the use of advanced 3D graphics was experimented in the visualization of temporal constructs expressed in the widespread formalism of Computation Tree Logic (CTL) [7]. However, no real time expressivity is provided, nor it can be added without severely saturating graphic representation. In [14], implications of sequencing properties expressed in *Temporal Logic of Actions* (TLA) are represented through state-transition diagrams with algebraic representation of actions. Also in this case, no real time constraints are encompassed in the formalism. Moreover, since diagrams do not completely cover TLA expressivity, they cannot replace textual representation of sentences.

In this paper, we introduce, motivate, and evaluate, a visual formalism for the complete and un-ambiguous presentation of a real time logic based on the temporal construct *until*. The logic, that we refer to as Temporal Logic of Traces (TLT), was developed as a part of a broader tool for computer aided verification [6], and has been equipped with a model checking algorithm supporting verification of sentences against the execution sequences of a densely timed transition system in the style of [5] and [10]. Expressive limits of TLT were defined by developing on syntactic and semantic characteristics of well consolidated real time logics, such as MITL and RTTL [3], and by adapting them so as to permit a semantically sound visual representation. The visual formalism supporting presentation of TLT sentences was designed joining heuristic design and user-based evaluation centered on the objective of supporting teaching of real time logics to the students of higher education courses in software engineering. Heuristic development was oriented to maximize consistency, i.e. to minimize the number of graphical metaphors used to map textual sentences to the visual representation. Individual metaphoric assumptions were defined by prototyping and exposing alternative graphical representations to a representative sample of the target community of expected users. The resulting notation has been implemented within an interactive syntax-directed editor which integrates the visual presentation with the conventional textual notation. Finally, the editor has been used to carry out a competitive user-based evaluation of the usability of textual and visual representations, by carrying out a readability test on a larger sample of potential end-users.

2 Temporal Logic of Traces

Temporal Logic has been widely addressed as a language to describe and reason about the temporal ordering of the actions of parallel and concurrent systems [21] [16]. Real time logics extend the expressive domain of temporal logics by introducing quantitative constraints on the time elapsed between actions.

In the Temporal Logic of Traces (TLT), sentences are referred to a system model represented as a sequence of states. Each state is characterized by a set of basic propositions capturing instantaneous operation conditions, and it is associated with a dense time interval capturing the time during which the system stays in the state. Basic propositions can be recursively composed through Boolean connectives, or through the temporal operator *until*. Dense-time constraints are added to the logic by introducing *freeze variables* in the style of [3]: the time at which a given condition is satisfied can be frozen into a clock variable, and multiple clock variables can be compared to express quantitative timing constraints. Syntax and semantics of TLT are formally captured in the rest of this section. A simple example is also reported in Fig. 1 to help intuitive understanding.

$$\psi = x.(q \wedge (p \cup (y.(q \wedge r) \ \& \ (y - x) \in [3,7])))$$

(a)

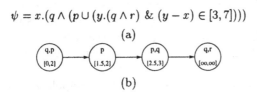

(b)

Fig. 1. (a) A temporal sentence expressed as a formula ψ of TLT: "in the initial state S, q is true (q) and, starting from S, p is true (p) at least until ($\cup$) reaching a state in which q and r are both true ($q \wedge r$) and the time elapsed from state S ($y - x$) lies in the interval [3,7] ($\in [3,7]$)". (b) A state sequence where ψ is satisfied.

Syntax
Given a set $\mathcal{P}$ of basic propositions, and a set V of clock variables, a TLT formula ψ is inductively defined as:

$$\psi := x_o.\phi\{ \ \& \ (x_o - x_i) \in [\alpha_i, \beta_i]\}$$
$$\phi := p \mid \neg\phi \mid \phi_1 \wedge \phi_2 \mid \psi_1 \cup \psi_2 \tag{1}$$

where $p \in \mathcal{P}$, $x_o, x_i \in V$ and $\alpha_i, \beta_i \in \mathbb{R}$, and where the curly brackets $\{,\}$ have the usual BNF meaning of zero-or-more repetition.

Semantics

Given a timed state sequence $\rho = (\sigma_i, I_i)$ made up of a sequence of states σ_i and time intervals $I_i \in \mathbb{R}^2$, and an interpretation (environment) ε $\varepsilon : V \mapsto \mathbb{R}$ assigning a real value to each clock variable, a TLT formula ψ is interpreted on ρ according to the satisfaction relation $(\rho, 0) \models_\varepsilon \psi$ (read " ρ satisfies ψ in the environment ε at time 0 ") which is defined by the following inductive clauses:

- $(\rho, t) \models_\varepsilon x_o.\phi\{x_o - x_i \in [\alpha_i, \beta_i]\}$ iff: (ρ, t) satisfies ϕ in the environment $\varepsilon[x_o := t]$ derived from ε by freezing the value t of the current time in the clock variable x_o; and any (optional) constraint $x_o - x_i \in [\alpha_i, \beta_i]$ on the difference between x_o and the previously frozen variables x_i is satisfied;
- $(\rho, t) \models_\varepsilon p$ iff the proposition p holds in the state σ_i when time t is in the interval I_i;
- $(\rho, t) \models_\varepsilon \neg\phi$ iff it is not true that $(\rho, t) \models_\varepsilon \phi$;
- $(\rho, t) \models_\varepsilon \phi_1 \wedge \phi_2$ iff $(\rho, t) \models_\varepsilon \phi_1$ and $(\rho, t) \models_\varepsilon \phi_2$;
- $(\rho, t) \models_\varepsilon \psi_1 \cup \psi_2$ iff $\exists t', t' > t$, such that $(\rho, t') \models_\varepsilon \psi_2$ and $\forall t'', t < t'' < t'$, $(\rho, t'') \models_\varepsilon \psi_2$.

With respect to [3], the two-level syntax of (1) introduces a syntactic constraint which imposes that clock variables can be frozen either at the beginning of the formula (i.e. the zero-time since which the entire formula is interpreted) or in correspondence of the two intermediate and final conditions of an until operator. This enforces the interpretation of each freeze variable as the time reference of an independent temporal context: while Boolean composition subtends simultaneity in the satisfaction of connected conditions, the temporal operator until locates intermediate and final conditions in subsequent states encountered along a temporal progression. While not reducing the expressive domain of satisfiable sentences, this limitation associates freeze variables with an intuitive physical meaning and facilitates the establishment of a semantically sound visual metaphor.

The two-level syntax of (1) also prevents the application of Boolean connectives to temporal contexts (i.e. it does not accept a formula such as $\neg\psi$). This limitation, which was introduced to simplify the problem of verifiability, reduces the expressive power of the logic by preventing the expression of quantitative constraints which impose a freeze variable to lay outside a time interval (i. e. it does not accept a constraint such as $y - x \notin [\alpha, \beta]$).

2.1 Shorthands

Further operators can be defined as shorthands by combining Boolean connectives and temporal operators. Boolean connectives such as *or* ($\vee$) and *implies* ($\rightarrow$), with their conventional meaning, are derived in the usual manner as $\phi_1 \vee \phi_2 = \neg(\neg\phi_1 \wedge \neg\phi_2)$ and $\phi_1 \rightarrow \phi_2 = \neg(\phi_1 \wedge (\neg\phi_2))$ respectively; the temporal construct $\Diamond\psi_2$ (read "eventually ψ_2 ") is defined in the conventional manner as $\Diamond\psi_2 = true \cup \psi_2$, which means that there exists a future state where ψ_2 holds; the construct $\Box\psi_1$ (read "always ψ_1 ") is expanded according to the syntactic reduction ψ_1 as: $\Box x_o.\phi_1\{x_o - x_i \in [\alpha_i, \beta_i]\} = \neg(true \cup (x_o.\neg\phi_1\{x_o - x_i \in [\alpha_i, \beta_i]\}))$, meaning that it is not the case that ψ_1 is false in some future state.

3 Mapping Text to Diagrams

In the design of a visual formalism for TLT, heuristic principles were applied to define a limited set of visualization metaphors supporting consistent representation of TLT basic semantic constructs. Alternative representations were defined where multiple choices were possible with no obvious expected preference. Heuristic assumptions were then refined and alternative representations were selected by resorting to the judgment of a representative sample of 15 expected end-users.

3.1 A Metaphoric Nucleus

TLT expressivity combines three distinct fragments: *(i)* temporal operators constraining the sequencing by which individual states are visited; *(ii)* quantitative constraints on the time elapsed between different states; and *(iii)* propositional logic capturing properties of individual states.

Temporal operators: Sequencing conditions expressed through temporal operators underlie the basic concept of a sequence of temporal contexts through which the system progresses. Following the metaphor of timeline graphs, this sequence is visualized as a horizontal path line (see Fig. 2a). According to semantic clauses of Sect.2, a path may be characterized by a terminal condition ψ_2 becoming true in the temporal context reached at the end of the path, and by a suspensive condition ψ_1 holding until then. This interpretation is visualized by annotating the path line of Fig. 2a with two boxes at the end and in the middle of the line, which represent the temporal contexts satisfying the terminal condition ψ_2 and the suspensive condition ψ_1, respectively (see Fig. 2b-c). As a consequence of these assumptions, the temporal progression defined by an until operator is represented as a path annotated with both the suspensive and the final conditions ψ_1 and ψ_2 (see Fig. 2d).

Fig. 2. Visualization primitives of temporal progression: (a) a path through the state transition system; (b) termination of a path on reaching of condition ψ_2 ; (c) satisfaction of condition ψ_1 along the states of the path; (d) a path where ψ_1 holds until reaching a state in which ψ_2 is satisfied.

Quantitative constraints: As mentioned in Sect.2, a freeze variable is the time reference of a temporal context characterized by some distinctive condition, which can be either the zero time at the root of the formula, or the time at which the system satisfies the suspensive or the final condition of a temporal progression expressed through the until operator. Therefore, each freeze variable can be identified either with the beginning of the statement, or with one of the boxes which enclose a suspensive or a final condition in the representation of Fig. 2. According to this, each freeze variable is represented by the box that encloses its represented temporal context (see Fig. 3a-b). Besides, optional constraints on the difference between two freeze variables are drawn as an optional set of arrow-ended lines connecting the boxes which represent the two contexts, and annotated with the constraint interval $[\alpha_i, \beta_i]$ (see Fig. 3c).

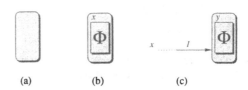

(a)   (b)   (c)

Fig. 3. Visualization primitives for quantitative constraints: (a) an independent temporal context; (b) freezing of the clock variable x to the temporal context of ϕ ; (c) constraint on the time elapsed from the temporal context identified by x (not shown in the figure) to the temporal context of ϕ, identified by variable y

Propositional Logic: Following a commonly accepted standard, basic propositions are represented in textual form within the frame of a text-sheet icon. For the representation of instantaneous conditions captured through Boolean combination, the assumption that the horizontal direction stands for time progression imposes that operands be arranged along the vertical direction. This rules out the possibility to express Boolean compositions using common visual conventions based on the spatial arrangement of operands, and rather advocates for an iconic representation of connective symbols. To this end, different set of icons are possible, including the conventional logic symbols $\neg, \wedge, \vee, \rightarrow$, the explicit text Not, And, Or, Imp, and the c-like notation !, &&, ||, − >, which can be expected to match the cultural background of a target population of students with a strong programming practice.

Acceptance of the three different styles was assessed by carrying out a test to evaluate the four usability factors of *learnability, efficiency, effectiveness*, and *satisfaction* [18] [19] [20]. In the test, users were first asked to select the set of symbols most conforming to their expectancy. Users were then requested to translate in natural language Boolean propositions expressed using the different connective symbols, and to indicate the notation which was easier to read. On completion of the reading test, users were asked to write some simple propositions.

Learnability was measured by the number of selections received by each notation (37% for c-like operators, 50% for logic symbols, 13% for explicit text) and by the number of times that the notation was used in the final writing test (24% for c-like operators, 66% for logic symbols, 10% for explicit text). No significative differences were evaluated either in the efficiency or in the effectiveness of different notations, as the time spent in carrying out the test was basically uniform and no errors were made by any user. Satisfaction was accounted as the number of times that each notation was indicated as to require the minimum effort (32% for c-like operators, 58% for logic symbols, 10% for explicit text). The representation selected according to these results is reported in Fig. 4.

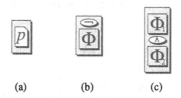

Fig. 4. Visualization primitives of Propositional Logic: (a) basic proposition p; (b) negation of a condition ϕ; and, (c) conjunction of conditions ϕ_1 and ϕ_2.

3.2 Visual Re-Writing Rules

Summarizing, the basic visualization primitives associated with the semantic constructs of temporal progression, timing constraints, and propositional logic, can be combined into a set of visual re-writing rules which associate each terminal token of TLT with a concrete drawing. Mimicking the BNF definition of (1), the visual re-writing rules for a generic formula ψ are expressed as:

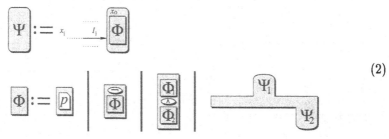

$$(2)$$

Matching the recursive organization of TLT syntax and semantics, these rules reduce the visualization of a formula to the recursive visualization of its subformulae and to the concrete drawing of graphic icons representing terminal symbols. While the structure of rules matches TLT syntax, concrete drawings that these rules produce reflect the semantics of terminal symbols.

As an example, Fig. 5 reports the visual representation for the textual formula of Fig. 1.

3.3 Visual Shorthands

As it occurs in the textual domain, shorthand visual representations are useful to manage notational complexity and reduce screen saturation.

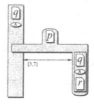

Fig. 5. Visual representation of the formula of Fig. 1.

For Boolean operators, shorthand binary connectives are represented with the same style as that of conjunction.

For the eventually operator ($\diamond\psi_2$), the visual representation of the shorthand expression (true $\cup \psi_2$) is that of Fig. 6a. Since the suspensive condition is **true**, the temporal context of the suspensive condition can be omitted, yielding the shorthand visual representation of Fig6b, which results to be fully consistent with the assumptions of the metaphoric nucleus of Sect.3.1. In the case that the temporal context of ψ_2 is subjected to quantitative constraints, the visual shorthand naturally evolves as reported in Fig. 6c-d.

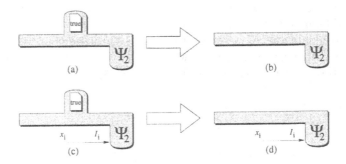

Fig. 6. Rational of the visual shorthand for eventually operator, without (a-b) and with (c-d) quantitative constraints.

For the always operator ($\square\psi_1$), in the case that the temporal context of ψ_1 is not bounded by a quantitative constraint, the visual representation of the shorthand expression ($\neg(true \; (\cup(x_o.\neg\phi_1))$) is that of Fig. 7a. In this case, there is not an obvious visual simplification and hence the visual shorthand must be defined by comparing the semantics of the always operator against the metaphoric assumptions of Fig. 2. This yields the simplified representation of Fig. 7b, which represents a path with the suspensive condition but without the terminating condition, in a dual manner with respect to the visual shorthand assumed for the eventually operator. This representation corresponds to the interpretation of $\square\psi_1$ as $\psi_1 \cup false$, which is intuitive but, alas, not formally correct with respect to the semantics of until. This clash between intuition and formal semantics emerges in the case that the temporal context of ψ_1 is bounded to a quantitative constraint, i.e. if the set of optional constraints in ψ_1 is not empty, as in Fig. 7c. In this case, the representation of Fig. 7b does not imply any obvious and consistent point where the constraint arrow-ended line can be placed. Figs.7d-e show the two simplest possible solutions which were devised by three concurrent designers. The representation of Fig. 7d represents the constraint as a bounding attribute of the temporal progression, but it looks more like a kind of until rather than an always construct. The representation of Fig. 7e overcomes this drawback, but it is not fully consistent with the metaphoric assumptions of Sect.3.1. In fact, in Fig. 3, the arrow-ended line denotes the time interval within which a condition becomes true, rather than an interval in which a condition is continuously satisfied.

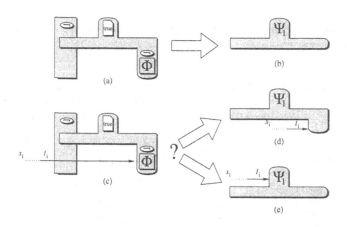

Fig. 7. Rational of the visual shorthand for always operator, without (a-b) and with (c-d-e) quantitative constraints. Two alternative representations are devised in the presence of constraints.

Since heuristic considerations do not yield a consequent design for the visual shorthand, the sample of end-users was asked again to provide feedback on the different solutions.

Correctness of heuristic assumptions, and acceptance of different alternatives was assessed by carrying out a readability test to evaluate *learnability* and *satisfaction* of the proposed visual notations. After introducing the underlying rational of the visual representation of until operator of Fig. 2d, users were requested to express in natural language the meaning of a few simple visual expressions involving temporal operators, with or without quantitative constraints. After each interpretation, users were asked to rank their acceptance by expressing whether the notation was "natural", "not immediate but consequent", "not consequent". Finally, on completion of the reading test, users were asked to write some simple propositions. Learnability was evaluated by the number of correct interpretations provided during the reading test (understandability), and by the number of times that the notation was used in the final writing test (memorability). Satisfaction was evaluated by computing a weighted average of the acceptance ranking.

Results of the test, which are summarized in Table 1, basically confirm the correctness of heuristic assumptions. As expected in the heuristic analysis, the visual representation of the always operator is a problem in the presence of quantitative constraints. With respect to this, it is worth noticing that a significant fraction of the users provided a common alternative interpretation for both the shorthand representations of the constrained always, which result to be basically consistent with the metaphoric nucleus of Sect.3.1: 36% of the users interpreted the representation of Fig. 7e as " ϕ_1 becomes true at a time instant within $[\alpha, \beta]$ and then holds indefinitely"; 54% of the users interpreted the representation of Fig. 7d as " ϕ_1 is initially true and holds until an undefined event occurs within $[\alpha, \beta]$".

According to these results, Fig. 7e was assumed as visual shorthand for constrained always.

Table 1. Results of user-based validation of visual representation of temporal operators. Two different representations have been tested for the constrained always, corresponding to the alternative shown in Fig. 7(d) and (e), respectively.

	Understandability	Memorability	Satisfaction $(0 \div 100)$
Eventually	82%		67
Always	73%		63
Until		100%	
Constr. Until	91%		80
Constr. Event.	100%	64%	82
Constr. Always (d)	36%	18%	75
Constr. Always (e)	54%	55%	50

4 Implementation and User-Based Evaluation of a Visual Syntax-Directed Editor

The visual representation of TLT sentences has been enforced within a syntax-directed editor [1].

By relying on their bijective syntactic relation, the editor supports integrated and coherent manipulation of sentences both in the visual and the textual notation. The two views are displayed in two separate frames (see Fig. 8) whose contents are maintained aligned by the system. While the visual representation facilitates most of the tasks involved in parsing and interpretation of TLT sentences, textual form fits the skill of experienced users anchored to conventional

---

[1] The editor has been developed in Java as a part of a CASE tool supporting reachability analysis and automated model checking of time dependent systems modeled as Time Petri Nets [6].

formal expressions. In an educational environment, the joint use of both the views is also useful to support teaching of the conventional textual notation.

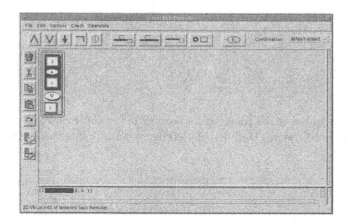

Fig. 8. A snapshot of the editor screen showing the visual form and the conventional textual notation jointly displayed in two separate frames on the top and bottom parts of the screen, respectively.

Both the views can be used in the interactive editing of expressions, following the pattern of conventional equation editors: statements are expressed by filling empty marks through the selection of operators and basic propositions from the icon toolbar on top of the window; in turn, operators introduce new empty marks which permit to recursively develop the formula in a top-down approach. Quantitative constraints are created by drawing an horizontal line between representations of different temporal contexts. Textual information related to basic propositions or quantitative constraints are entered by conventional dialog boxes.

Icons for conventional editing commands are grouped in a toolbar on the left of the window. In particular, commands include a special-purpose folding/unfolding command, and the conventional cut & paste. Folding/unfolding permits to implode/explode a sub-formula to manage screen saturation through detail hiding (see Fig. 9). Cut & paste uses a clipboard whose contents can be explicitly displayed in the right part of the screen, in both their visual and textual form, so as to permit selective paste of sub-formulae in the clipboard (see Fig. 10).

Customization of visual parameters (e.g. length, proportion, color of path and box drawings, font dimension, etc.) is supported through the invocation of the "configuration" menu. Interfacing with the verification model checking tool is available through the "verification" menu.

Other conventional features of the interface are not described here.

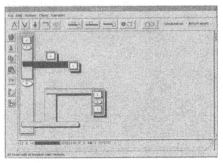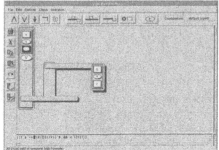

Fig. 9. A snapshot of the editor screen showing the use of the folding/unfolding command. The sub-formula selected in the left figure is folded into the selected box of the right figure.

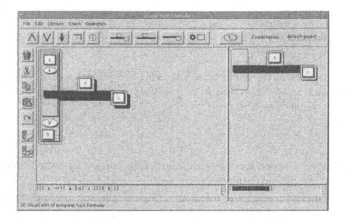

Fig. 10. A snapshot of the editor screen showing the explicit clipboard. The right frames show the contents of the clipboard after the copy of the selected sub-formula in the left frames.

4.1 Textual versus Visual Representation

The possibility to let the user edit sentences both in the textual and visual form within a common environment, opens the way to a competitive evaluation of the usability of the two notations.

A user-based test to accomplish one such evaluation has been designed which addresses the capability of the two notations to support parsing and interpretation of TLT sentences. The test has been administered to a pilot sample of 6 users to refine the contents and evaluate the expected duration, and then has been iterated on 15 students attending the current course in software engineering at school of engineering of the University of Florence. Students attending the course are at the end of a higher education curriculum in computer engineering with a strong content in mathematics, engineering, and computer programming, but with a limited education in logics.

4.2   Test Plan and Pilot Results

Before the test session, users attended two introductory lectures on the use of real time logics in the verification of time-dependent systems, and on the specific characteristics of TLT. Users were shown both the standard textual notation and the proposed visual representation, and they attended a demo on the use of the syntax-directed editor.

Each user performed two stages of testing to experience both the visual and the textual notation. To this end, users were organized in two groups testing the two notations in opposite orders (i.e. before visual and then textual or viceversa). During each stage, users were shown a set of tests sentences of TLT, with different and increasing complexities. Each set included: *1)* elementary expressions with a single operator, aimed at recalling representation assumptions; *2)* Boolean expressions (with no temporal operators) with 2 different levels of complexity; *3)* temporal expressions, with no quantitative constraints, with 3 different levels of complexity; *4)* temporal expressions with quantitative constraints with 2 different levels of complexity. For each test sentence the user was requested to accomplish two subsequent tasks: the *reading task* consisted in writing the meaning of the sentence in natural language; the *deduction task* consisted in deciding the validity of some implications of the test sentence, which were designed so as to require understanding and reasoning about the sentence meaning. After completion of both the visual and the textual testing stages, the user was shown back all the sentences, with both the textual and visual representation, and was asked to subjectively rank the different level of effort required by the two notations, for each sentence and for each of the two testing tasks. Tests regarding the two most complex sentences in the set are reported in Fig. 11; a complete report of the entire test session does not fit in the space of this paper, but it is available as http://aguirre.ing.unifi.it/~lusini/dafne/tests.ps.

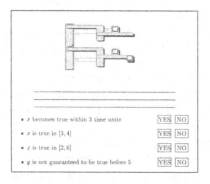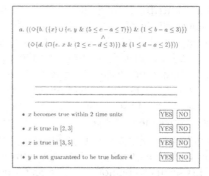

Fig. 11. An example of the style of sentences and questions proposed to the user during the test session.

During the test session, the following measurements were taken: learnability was measured by the time spent and the number of errors made in the translation task; efficiency and effectiveness were measured by the time spent and the number of errors made in the deduction task; satisfaction was measured by the subjective effort ranked in the post-test questionnaire. Results are summarized in Table 2 and they are discussed in the following of this section.

Table 2. Competitive results obtained for the visual and textual notation in the reading and deduction tasks for each category of sentences. Ranked effort is expressed as the degree of preference for the visual notation in a scale from 0 (text is definitely better) to 100 (graphics is definitely better).

	Translation					Deduction				
	Error Rate (%)		Time (sec)		Visual	Error Rate (%)		Time (sec)		Visual
	Textual	Visual	Textual	Visual	Pref.	Textual	Visual	Textual	Visual	Pref.
Elem.	1.7%	1.7%	24	24	37.9	4.8%	8.1%	12	12	46.5
Bool.	19.8%	19.4%	41	52	25.5	12.0%	18.6%	49	41	27.3
Temp.	6.1%	6.1%	53	66	57.6	37.1%	19.2%	80	43	75.8
Const.	65.2%	35.5%	133	108	93.6	48.9%	23.8%	123	63	94.5

Results of the test sessions basically confirmed the validity of the planned battery of tests. In particular, according to measured error ratios, the complexity of selected statements and the preliminary user training were proven appropriate to discriminate usability of the two representations. The overall test lasted an average time of 32 minutes plus the post-test questionnaire and discussion, which also appears appropriate to undertake a more extensive experimentation. Regularity of results obtained permits to draw some conclusions.

For Boolean expressions, conventional text performs better than the visual notation, both in the time spent and in the subjective effort ranked. This appears not only in the case of Boolean compositions, but also from the analysis of results obtained with mixed sentences in which Boolean and temporal operators contribute differently to the overall sentence complexity. This was somehow expected: on the one hand, the visual representation of Boolean compositions does not provide any metaphoric support to the interpretation; on the other hand, for syntactic parsing, text is more efficient in terms of the ink-ratio score [24], and more fitting to the conventional horizontal arrangement of sentences to which users are accustomed. Actually, the vertical arrangement of Boolean compositions (and, consequently, the need to represent them in graphical form) is a consequence of the allocation of the horizontal axis to the representation of the temporal dimension: allocating time on the vertical axis could overcome the limit, but it would violate the conventional metaphor of horizontal timeliness. This seems to be not effective as temporal compositions are the prevailing factor of complexity in both learning and usage of TLT.

In the reading task, text performs better than graphics in the time spent and in the ranked effort, provided that the sentence does not include quantitative constraints. In fact, while the textual form can be cast in natural language with a one-to-one syntactic transposition of individual textual symbols, the 2D arrangement of visual expressions needs to be re-sequentialized, and forces the user to organize and partially interpret the sentence. This results in a longer time spent and a larger ranked effort, but actually corresponds to a different and more fruitful user's activity. This is confirmed by the shorter time, the lower ranked effort, and the more accurate results, obtained in the accomplishment of deductive tasks, which actually comprise the real final goal of the user in his/her real activity. This advantage of the visual form appears and grows when complexity of statements increases: while text is appropriate for very simple expressions, the visual form can become a powerful tool to scale the affordable level of expressive complexity.

In the presence of quantitative constraints, the visual notation definitely overperforms text in all measured figures. With respect to this, it is worth noting that, while its textual representation may appear overwhelming, also the most complex statement of Fig. 11 corresponds to a quite simple requirement, which may easily occur in the characterization of a real time system. A measure which does not appear from the figures in Table 2, is the quality of results provided in the translation task. While translations from text closely matched the syntactic organization of the formal sentence, with no added value, those derived from the visual form were able to capture the actual meaning of expressions. This confirms that, in the reading task, the visual notation stimulates and facilitates synthesis and understanding.

References

1. J.F.Allen "Maintaining Knowledge about Temporal Intervals," *Communications of the ACM*, Vol.26, No.11, 1983.
2. R.Alur, C.Courcoubetis, D.Dill, "Model-Checking for Real-Time Systems," *Proc. 5th Symp. on Logic in Computer Science*, Philadelphia, Jun.1990.
3. R.Alur and T.A.Henzinger, "Logics and Models of Real Time: A Survey," *Real-Time: Theory in Practice*, J.W. de Bakker, K. Huinzing, W.P.de-Roever, G. Rozenberg (Eds), *Lecture Notes in Computer Science* 600, Springer Verlag, 1992.
4. R.Alur and T.A.Henzinger, "Real-Time Logics: Complexity and Expressiveness," *Information and Computation*, 104(1), 1993.
5. B.Berthomieu, and M.Diaz, "Modeling and Verification of Time Dependent Systems Using Time Petri Nets," *IEEE Transactions on Software Engineering*, Vol.17, No.3, Mar.1991.
6. G.Bucci, E.Vicario, "Compositional Validation of Time-Dependent Systems Using Communicating Time Petri Nets," *IEEE Transactions on Software Engineering*, Vol.21, No.12, Dec.1995.
7. E.M.Clarke, E.A.Emerson, A.P.Sistla, "Automatic Verification of Finite-State Concurrent Systems Using Temporal Logic Specifications," *ACM Transactions on Programming Languages and Systems*, Vol.8, No.2, Apr.1986.
8. B.Dasarathy, "Timing Constraints of Real-Time Systems: Constructs for Expressing Them, Methods for Validating Them," *IEEE Transactions on Software Engineering*, Vol.11, No.1, Jan.1985.
9. A.DelBimbo, L.Rella and E.Vicario, "Visual Specification of Branching Time Temporal Logic," *Proc.Symp on Visual Languages*, VL95, Darmstadt, IEEE Comp. Soc. Press, 1995.

10. D.Dill, "Timing Assumptions and Verification of Finite State Concurrent Systems," *Proc. Workshop on Computer Aided Verification Methods for Finite State Systems,* Grenoble, France, 1989.

11. D.Harel, "On Visual Formalism," *Communications of the ACM,* Vol.31, No.5, May.1988.

12. A.M.Haber, Y.E.Ioannidis, M.Livny, "Foundations of Visual Metaphors for Schema Display," *Journal of Intelligent Information Systems,* Vol.3, Summer 1994.

13. G.Kutty, Y.S.Ramakrishna, L.E.Moser, L.K.Dillon, P.M.Melliar Smith, "A Graphical Interval Logic Toolset for Verifying Concurrent Systems," *Proc. of the fourth Workshop on Computer Aided Verification,* Montreal, Canada, Jun.1992.

14. L.Lamport, "TLA in Pictures," *IEEE Transactions on Software Engineering,* Vol.21, No.9, Sep.1995.

15. G.Kutty, L.K.Dillon, L.E.Moser, P.M.Melliar Smith, Y.S.Ramakrishna, "Visual Tools for Temporal reasoning," *Proc. VL'93,* Bergen, Norway, 1993.

16. Z.Manna, A.Pnueli, "The Temporal Logic of Reactive and Concurrent Systems," Springer Verlag, New York, 1992.

17. L. Moser, Y. S. Ramakrishna, G. Kutty, P. M. Melliar-Smith and L. K. Dillon, "A Graphical Environment for Design of Concurrent Real-Time Systems", *ACM Transactions on Software Engineering and Methodology,* Jan.1997.

18. J.Nielsen, "Usability Engineering," Academic Press, San Diego, Calif. 1993.

19. J.Nielsen, "Iterative User Interface Design," *IEEE Computer,* Nov.1993.

20. J.Nielsen, "The Usability Engineering Life Cycle," *IEEE Computer,* Mar.1992.

21. A.Pnueli, "The Temporal Logic of Program," *Proc. 18th Annual Symposium on Foundations of Computer Science,* IEEE Comp. Soc. Press, 1977.

22. J.A.Stankovic, K.Ramamritham, "What is Predictability for Real Time Systems," *Journal of Real Time Systems,* Vol.2, Dec.1990.

23. J.A.Stankovic, "Misconceptions About Real Time Computing," *IEEE Computer,* Vol.21, No.10, Oct.1988.

24. E.R.Tufte, "Envisioning Information," Graphic Press, Cheshire (CT), 1990.

25. J.Xu, D.Parnas, "On Satisfying Timing Constraints in Hard Real Time Systems," *IEEE Transactions on Software Engineering,* Vol.19, No.1, Jan.1993.

Visualising the Behaviour of
Intelligent Networks

Carla Capellmann[1], Søren Christensen[2], and Uwe Herzog[1]

[1] Deutsche Telekom, Technologiezentrum
D-64307 Darmstadt, Germany
E-mail: {capellmann, herzog}@tzd.telekom.de
Telephone: +49 6151 83 3070 Fax: +49 6151 83 4221
[2] Computer Science Department, University of Aarhus,
Ny Munkegade 116, DK-8000 Århus C, Denmark
E-mail: schristensen@daimi.aau.dk
Telephone: +45 8942 3188 Fax: +45 8942 3255

Abstract. Telecommunication systems in general, and Intelligent Networks as well, are large distributed systems with a high degree of concurrency. To investigate the specific aspects of the behaviour of such systems adequate formal specification techniques are required.

In this paper, we show how designers of services in an Intelligent Network can use Coloured Petri Nets, to obtain a description of the services which on the one hand is precise and unambiguous, and on the other hand provides the basis for a comprehensive and thorough understanding of the behaviour by visualising the processes. Coloured Petri Nets supported by adequate tools can therefore serve as basis for investigations of new services to be provide.

For a formal description technique to be successful, it is necessary to have tool support. This provides the designer with support for the development and test of the design. Further more the tool should also allow flexible graphical feedback which can be tailored to the needs of the specific domain in question.

In this paper we show how the overall system behaviour of an Intelligent Network can be visualised, how the user can interact with the system, and how the history of important events can be captured in ways which are easy to understand for engineers familiar with Intelligent Networks, even though unfamiliar with the formal description technique in use.

1 Introduction

The ability to investigate how the most required Intelligent Network services can be produced at reasonable costs is a crucial one, a formal specification technique is required and it allows for simulation and analysis of its behaviour in relation to different services. The standard for Intelligent Networks (IN) [10] contains an SDL [8] specification of the Basic Call State Model (BCSM) [12], which was, however, unsuited for our purposes. Since Petri nets, especially high-level Petri nets, are very well suited to specify distributed and concurrent systems and have proved their applicability also in the area of telecommunications [1–3], we developed a formal specification of the BCSM using Coloured Petri nets [7].

Our objectives were to obtain a specification of the BCSM which:

1. was precise and unambiguous,
2. would provide a comprehensive and thorough understanding of the BCSM behaviour by visualising the processes within the BCSM, as well as how BCSM and IN service logic interact during IN service execution, and
3. would therefore serve as basis for investigations on service provisioning.

By use of formal specifications, we wanted to clarify issues left open by the standards. Furthermore, the formal specifications should allow for simulation of the behaviour, thus visualising the BCSM's dynamics.

In section 2 we introduces the concepts of Intelligent Networks which are needed in the rest of the paper. Section 3 presents Coloured Petri Nets and discusses the kind of visualisation that it directly supports. In section 4, we present other visualisations of the dynamics of basic call processing in Intelligent Networks, also based on the formal specification. Finally, section 5 concludes the paper.

2 Intelligent Networks

Today, telecommunication networks are faced with the new requirement that they should provide the ability to collect and compute information in addition to the classical transmission and switching functions, i.e., to provide network intelligence. To this end, the Intelligent Network [5, 18] is recognised as one of the most important evolutionary developments these days. The IN is understood as a concept for the fast and economical creation and provision of new telecommunication services that go beyond the limits of conventional telephony.

IN is applied to switched telephone networks (Public Service Telephone Network and Integrated Services Digital Network). The IN networks were introduced in the mid 80's in the US. In 1988, IN standardisation started and today the Intelligent Networks are standardised by the International Telecommunication Union [10]. Enhancements and adaptations for the European market are specified by European Telecommunications Standards Institute. Since this time, IN has been introduced in telecommunication networks world-wide. Examples of IN services are Freephone, Card Calling Services, and Premium Rate Services.

2.1 Architecture of Intelligent Networks

The basic idea underlying the IN architecture is the separation of basic switching and transport functionality on the one hand and service logic functionality on the other. Instead of replicating the service control logic in all telephone exchanges, it is centralised in specific network nodes. From these nodes, the switches can be controlled remotely using the Signalling System No 7 network.

The IN is a functional-oriented architecture. Its logical architecture clearly identifies the different types of functionalities and represents them in terms of functional entities. In Figure 1, we show how these functional entities are mapped onto physical network elements, allowing a more flexible adaptation of the physical network configuration to changing market requirements.

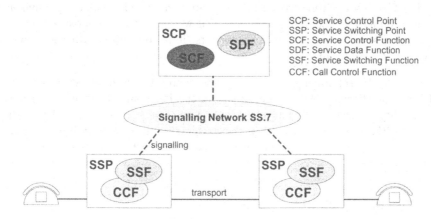

Fig. 1. Major Components of an IN architecture

The functional entities comprise:

Call Control Function: The Call Control Function (CCF) represents the switching functionality contained in any local exchange that is not yet prepared for handling IN calls.

Service Switching Function: The Service Switching Function (SSF) is always at the same location as the Call Control Function, enhancing it with the capability to handle IN calls. The Service Switching Function recognises the requests for IN services and sets up and sustains the communication relationship with the functional entity containing the IN service logic.

The Call Control Function/Service Switching Function is physically located in the Service Switching Point (SSP), which is an IN-capable local exchange.

Service Control Function: The Service Control Function (SCF) contains the service logic for all IN services. It is located in the Service Control Point (SCP).

Service Data Function: The Service Data Function (SDF) contains the database for the IN network. It contains service-specific and user-specific data (User profiles etc.). Also charging records can be stored in the Service Data Function. It can be located in the Service Data Point (SDP), and, as in most current products, also in the Service Control Point.

Other functional entities which are part of an IN are briefly listed below. For our purpose, however, they are unimportant and are thus not considered in the subsequent sections of this paper.

Special Resource Function: The Special Resource Function models any service-specific resource (bridge, digit collector, ...); it is connected with the Intelligent Peripheral network element.

Service Management Function: The Service Management Function models the management functions; it is associated to the Service Management System.

Service Creation Environment Function: The Service Creation Environment Function models the creation of new services by composing them of a set of service-independent components; it is connected with the Service Creation Environment.

Note that by separating the service logic related functionality (Service Control Function, Service Data Function) from the basic switching and transport functionality (Call Control Function), new services can be introduced into the network without having to modify the exchanges (Service Switching Points). Thus, their provisioning becomes more time- and cost-efficient, which is the overall goal of IN.

2.2 The Basic Call State Model

In this context, one important issue is the "interface" between service control logic and the underlying network. The service logic must be able to modify the basic call processing in order to realise an IN service, i.e., a service that goes beyond basic telephony.

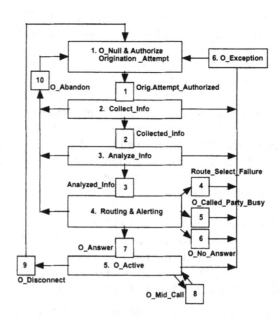

Fig. 2. Originating BCSM of IN Capability Set 1

In the IN architecture, the switching and message transport functions of the conventional telephone network are represented by the Basic Call State Model (BCSM). The BCSM defines the interface between the IN service logic and the underlying network. Its structure and granularity influence the kind of services that can be realised and in which way.

The BCSM describes the basic call processing in terms of states and transitions between them. It consists of two finite state machines: one, called Originating BCSM (see Figure. 2), describing the switching and transmission functions of the conventional telephone network at the originating party, and the other, called Terminating BCSM, representing the mentioned functions at the terminating party.

A state in the BCSM is referred to as a Point in Call, shown as big rectangles in Figure 2. The Points in Call describe the complete process of a call from picking up the receiver until the receiver is put down: in the Originating BCSM for the originating party, and in the Terminating BCSM for the terminating party. In addition to these Points in Call, there are points in the BCSM that are referred to as Detection Points, shown as small square boxes in Figure 2.

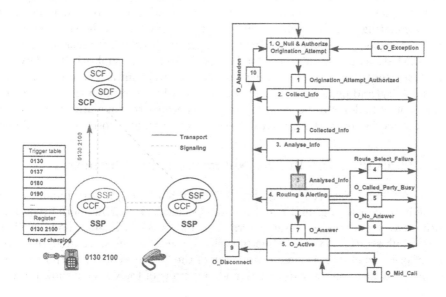

Fig. 3. Example of IN service execution

From these Detection Points, the IN service logic can be invoked if the required conditions for an invocation are fulfilled. As the service logic is not a part of the BCSM, it is not included in Figure 2. After the execution of service logic, the processing in the BCSM continues at the invoking Detection Point or another; which Detection Point is returned to depends on the requested IN service.

Consider a typical IN service, e.g., Freephone. Here the basic call processing starts in the Originating BCSM and proceeds until Detection Point 3, where it is recognised that this call is an IN call (see Figure 3). For every call, this Detection Point checks if the dialled number starts with an IN prefix, 0130 in the case of Freephone in Germany. In case of a Freephone call, the service logic is invoked and charging is modified so that the callee (terminating party) will be charged instead of the caller (originating party). The service logic will determine the concrete physical address to which the Freephone call is to be routed and return this number to the Service Switching Point. Call processing is resumed. The call will now be processed as every telephone call, with the exception that the Service Control Point will be notified of start and end of the connection in order to handle charging.

2.3 Motivation and Objectives

As the BCSM defines the points in basic call processing at which service logic can be invoked and modified, its structure and granularity have a major impact on the kind of services that can be realised in the IN, and in which ways. As a consequence, it is important to find the "right" structure and granularity of Originating and Terminating BCSM: If the structure of the specification is too coarse, only a limited amount of simple services will be feasible. A very detailed structure would allow for lots of different and sophisticated services, but tremendous costs for adapting the basic network infrastructure with a big complexity, and higher failure probability would be the consequence. Therefore, the issue of building a specification with an appropriate granularity that enables the production of most required IN services at reasonable costs is a crucial one.

3 Modelling Intelligent Networks using CP-nets

Coloured Petri Nets (CP-nets) [13, 14] is a formal description technique which combines the use of graphics and textual specification. Furthermore it contains primitives to decompose large descriptions into a set of related modules. This section illustrates how the CP-net describing the BCSM is structured and gives an impression on the size and complexity of the CP-net. For the purpose of this paper, it is not necessary to understand all details of the CP-nets shown in this section.

In the project we have used the tool Design/CPN [15] to support the use of CP-nets. We have used Design/CPN for editing, syntax check, and simulation. All figures shown in the rest of the paper are taken directly from the tool.

3.1 Overview of the CP-net

Besides the parts specifying the BCSM, the CP-net specification of IN consists of two more components: one representing the user and one representing the service logic. Figure 4 gives an overview of the overall CP-net. The rounded boxes represent CP-net modules and the arcs define the refinement relationship.

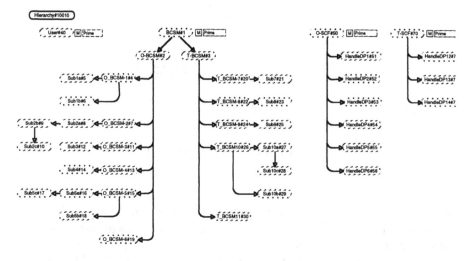

Fig. 4. Overview of the modules of the CP-net

3.2 A CP-net Description of Intelligent Networks

During the creation of the CP-net, two important requirements had to be taken into account: Firstly, the CP-net and its behaviour should be easily understandable so that it could be used to demonstrate to other people in the IN area, who are not familiar with CP-nets, what kind of processes take place in the BCSM and how BCSM and service logic interact when an IN service is accomplished. Secondly, the graphical appearance of the CP-net should be similar, or at least similarly structured as the BCSM, known in the "IN world" (see Figure 2).

The basic idea of a CP-net is that you represent states by means of places, usually drawn as ellipses, and actions by means of transitions, usually drawn as boxes. Places contain tokens having a value which belongs to the type of the place. Places and transitions can be connected with arcs, and an arc towards a transition specifies a precondition while arcs towards a place specify an effect of the occurrence of a transition.

The top level of the CP-net consists of the following components (see Figure 5): two hierarchical transitions, representing Originating and Terminating BCSM respectively, and two places (O-BCSM Input and T-BCSM Input) connecting them and representing the communication between them. The place O-BCSM Input has the type NumXMsgXD which is defined to be a triple consisting of a number identifying the phone, a message to be communicated, and the identity of the phone sending the message. The communication is defined in the IN standard as well, where it is graphically represented by dotted arrows between the two parts of the BCSM, this was however not included in Figure 2.

The two hierarchical transitions are refined by CP-nets. The CP-net refining the Originating BCSM is shown in Figure 6. It represents each Point in Call and its succeeding Detection Points as hierarchical transitions and relates them by

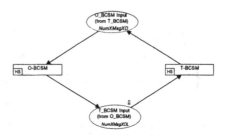

Fig. 5. CP-net for BCSM (top level)

places according to the transitions defined in the BCSM. In addition, the places are connected to those hierarchical transitions where messages are exchanged between the two BCSM sides.

It is instructive to compare the CP-net for the Originating BCSM (Figure 6) with its representation in the standard (Figure 2), the structure of the CP-net is very similar to the standard model, which was a major requirement to the formal specification.

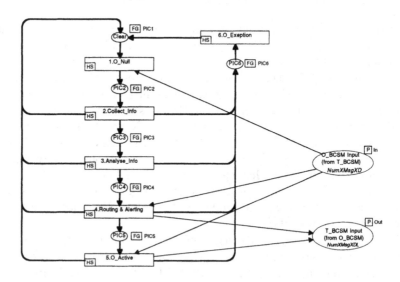

Fig. 6. CP-net for Originating BCSM

Each of the hierarchical transitions on this level are refined in turn. The CP-nets on the next lower level specify the relation between one Point in Call, its succeeding Detection Points and its "environment" (service logic, other BCSM side, user). The design principles for Points in Call and Detection Points were basically as follows:

- Each Point in Call is represented by a hierarchical transition with one incoming place. The actions carried out in the Point in Call are specified by the refining CP-nets.
- Each Detection Point is represented as a transition related to one incoming place and two outgoing places. This is done since a Detection Point represents a choice: call processing either continues in the BCSM (Detection Point not armed) or it is checked whether to invoke service logic (Detection Point armed).

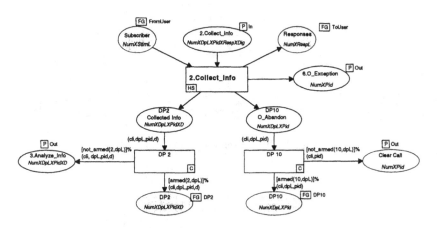

Fig. 7. CP-net for Point in Call 2, Detection Point 2, Detection Point 10

Figure 7 shows the CP-net for Point in Call Collect_Info which is followed by Detection Point 2 and Detection Point 10. Figure 8 presents the CP-net specifying the actions that take place during the collection of a sequence of digits dialled by the caller. While the Figures 5, 6, and 7 mainly show how the description can be decomposed, Figure 8 illustrates how the detailed behaviour of the system is specified. For all arcs you find expressions which specify the tokens to be moved along the arc. These expressions are written in a functional language called CPN ML, which is a variation of Standard ML [6].

3.3 Simulation a CP-net

In the Design/CPN tool it is possible to simulate the CP-nets using the graphics of the CP-nets in a very direct way. The state information is shown directly on the places, i.e., in Figure 9 the place Subscriber carries the information that the phone 3070 has dialled the digit "7". And the enabling of a transition is indicated by the fat dotted border. This means that the transition "Tone_off when 1st Stimuli recognise..." is ready to occur.

Working with interactive simulations on this level is very much like debugging a complex program, i.e., you work directly in the concepts of your specification. It

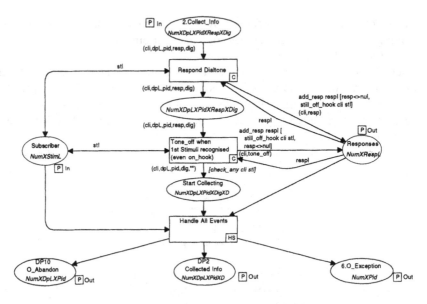

Fig. 8. CP-net refining Point in Call 2

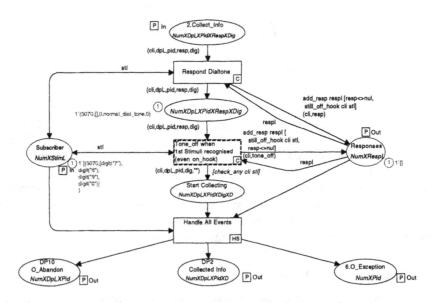

Fig. 9. Interactive Simulation of a CP-net

is also possible to make fast automatic simulations where the simulator executes a sequence of transitions until some stop criteria is met, and then the graphics is updated and can be inspected by the user.

3.4 Assessment of the Specification

Reconsidering the objectives we wanted to meet with this specification, we soon realised that we could achieve them only partially. We certainly obtained a specification that is precise and unambiguous. Details missing in the standard or issues left open have been identified during the modelling process and resolved in the resulting specification. The possibility of tool supported simulation by playing the token game has proved to be a very valuable help to designers. The complex behaviour of the specification would, otherwise, have been hardly comprehensible. By allowing different modes of simulations (interactive vs. automatic), the designer can test the overall behaviour in certain situations, examine specific parts, etc., and thereby gain confidence in the "correctness" of the specification.

Although a token game simulation visualises the dynamics of the specification, there are two shortcomings with this kind of visualisation:

- The size and complexity of the specification make it hard to keep an overview when investigating its behaviour. Remember that we want to investigate questions like the optimal BCSM granularity. Examining the CP-net by playing the token game can be rather tedious.
- Token game simulation requires knowledge of CP-nets. In the area of IN, most people are unfamiliar with CP-nets — this was one of the reasons to design the CP-net to structurally resemble the standard model as much as possible. Even though we did achieve this latter goal, the token game simulation does not help IN people understand the processes within the BCSM and its interaction with IN service logic.

To overcome these problems, we decided to add other kinds of visualisation to our CP-net. The basic idea was to visualise the behaviour in terms of the IN, i.e., using the domain-specific concepts. The different extensions are presented in the next section.

4 Visualisation

As our major objective is to visualise the behaviour that can take place in the BCSM during IN service execution as well as the interaction between the two IN architectural components this could not be achieved by one type of animation. Hence we developed different visualisations. Each of them addresses one specific aspect of visualising the behaviour of INs. Together, they complement each other, and, depending on the respective goals, one can select one or the other (or both).

All visualisations are based on the CP-net presented in the previous section. This was done by adding code segments to certain transitions. These code segments are executed whenever the transition occurs. Design/CPN offers libraries [4, 17] with different graphical animation functions that can be used for this purpose. It is important to notice that the behaviour of the formal specification is not in any way affected by the additional graphical visualisation added, and that new visualisations can be added independently of the other means of feedback.

4.1 Getting Overview Information

Aiming at a visualisation of the behaviour that takes place during basic call processing, one important aspect to show during a simulation is the actual status of call processing. As the BCSM (in form of its state machine representation, see Figure 2) is well-known in the IN world, these two diagrams are used as a basis. During a simulation, the Points in Call and/or Detection Points that are currently active are highlighted. Figure 10 shows an example where a connection between two parties/users has been established, i.e., one phone in the O_Active Point in Call and one in the T_Active Point in Call.

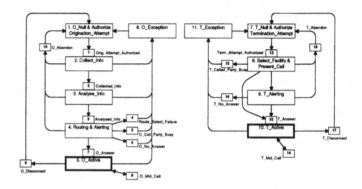

Fig. 10. Snapshot of BCSM overview diagram

By means of this graphic we can see the actual status of the BCSM at any time during the simulation. The information shown is only a part of the information of the system behaviour; it is, however, very useful as it provides an overview on the actual BCSM state. What can be observed is how the active phones progress in terms of Points in Call and how the Detection Points are tested as the system progresses.

4.2 Supporting User Interaction

In order to control the simulation on a more abstract level, i.e., not directly on the CP-net level, a second interface to the CP-net was added that enables the simulation user to interact with the simulation. To this end, a small IN is represented consisting of one Service Control Point connected to two Service Switching Points, each of which has three phones attached to it (see Figure 11).

After having started a simulation, the simulation user can select a graphical object. Depending on the state of the system, different kinds of interaction are possible: if a phone is on-hook, it is possible to select the phone which will result in changing its state to off-hook. In this state, the simulation user can select either the phone or its receiver. Selecting the receiver corresponds to an on-hook operation. Selecting the phone will prompt the simulation user for digits.

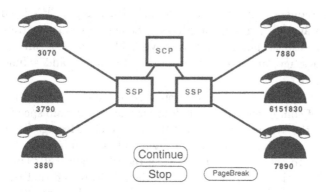

Fig. 11. Initial state of User Interaction Graphics

Figure 12 shows how the user interaction graphics look in the state where the two phones 3070 and 7890 are active and have established a connection. In this case, the user can choose either to hang-up one of the active phones, or to lift the receiver of one of the passive phones.

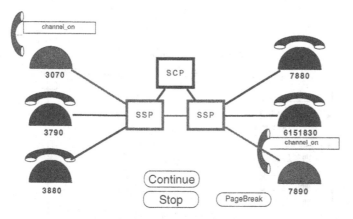

Fig. 12. Snapshot of User Interaction Graphics

This visualisation is used for controlling the simulation run and testing different scenarios, and it allows users which are unfamiliar with the CP-nets to test the behaviour of the system.

4.3 Capturing the Evolution of the System

It is often convenient to collect information on important events of the system. One way of doing this is to generate Message Sequence Charts [9]. The use of Message Sequence Charts in connection with Design/CPN is supported by a special library [4]. Adding the Message Sequence Chart is straight-forward; for each transition sending a message, we add a function call to display the respective message, and for each event of interest, we add a function call to update the chart.

Message Sequence Charts are often used to specify normal and/or exceptional cases of communication, and often serving as the main specification of the behaviour of the systems. In contrast to this, we use Message Sequence Charts to capture a single simulation run. In Figure 13, we can inspect the sequence of events needed to get to the situation with the two phones 3070 and 7890 being connected. Note that the system starts in the idle state and that the state of the system at the end of the Message Sequence Chart is the same state as depicted in Figures 10 and 12. The first event in the Message Sequence Chart is off-hook for phone 3070, this is immediately acknowledged by a Normal_dial_tone. Then there is an input of the digits for calling the phone 7890, and when the first digit arrives this is marked in the process called Origination Service Switching Point. This event corresponds to the occurrence of the transition depicted in Figure 9. After all digits have been collected the Originating Service Switching Point sends a connection request (Con_req) of phone 7890 to the Terminating Service Switching Point, which then sends a message to make the phone 7890 start ringing, and informs the Originating Service Switching Point that the phone 7890 has been alerted.

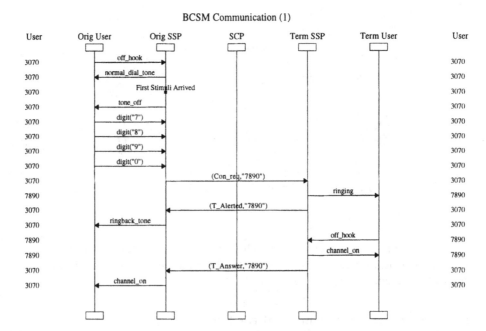

Fig. 13. Message Sequence Chart

It is important to notice that all of the events shown in the Message Sequence Chart are easy to interpret and understand when you have knowledge of the IN standard.

4.4 Future Extensions

It is possible to add further extensions of the visual feedback and we plan to do this in the future. One example could be to add visual feedback for the services of the IN, e.g., having bar-charts illustrating the state of a set of calls to a tele-voting service. Or to extend the graphics to handle more that one pair of Originating BCSM/Terminating BCSM communication pairs, e.g., to discuss different implementations of call-forwarding.

It is our impression that the model scales quite well, it is relatively easy to include even complicates services as long as they rely on the basic facilities included in the model. This means that it is not possible to discuss charging, or to test performance of the system since neither of these aspects are included in the basic model.

5 Conclusions

In this paper, we presented an approach towards visualising the behaviour of Intelligent Networks. It was based on a formal specification of the BCSM. As specification technique we have used CP-nets. The CP-net has been extended to visualise the model's behaviour in terms of IN specific concepts. Though all visualisations were based on the CP-net specifying the BCSM behaviour, no knowledge of CP-nets is required during a simulation. In this way, we benefit from formalisation (precise, unambiguous and executable specifications) without having to cope with its disadvantages (hard to understand for non-experts). This is an important step towards an application of formal description techniques in practice.

By visualising the system's behaviour in domain-specific concepts, we were able to overcome two problems: Users of the simulation, usually experts in the domain of IN or telecommunications, do not need to understand the formal description technique, but can still gain from its advantages. This may help convincing them about the general usefulness of formal specifications. Secondly, the complexity of the underlying specification can be hidden, enabling simulation users to focus on domain-specific questions. As a consequence, testing gets easier. One should be aware though, that visualisation does not make the task of specification any easier.

Though it seems obvious that visualisation should be made in terms of domain-specific concepts, this task is not an easy one. The question on what information to include and how to present it cannot be answered in general. Often, it depends on the specific purpose for which the visualisation shall be used. In our example of IN, a visualisation, as we have presented, may differ from one used to investigate service interaction problems. Both can however be built on top of the same formal specification.

Since there seems to be no "silver bullet", i.e., one visualisation for all purposes, it is very important to have a tool with related libraries supporting all kinds of different graphical presentations which can easily be integrated into the formal specification, e.g., Design/CPN.

References

[1] Capellmann, C.; Dibold, H.: Formal Specifications of Services in an Intelligent Network using High-Level Petri Nets. In Proceedings of the Case Studies Tutorial/Petri Nets'94, Zaragoza/Spain, June 1994

[2] Capellmann, C.; Dibold, H.: Petri Net based Specifications of Services in an Intelligent Network — Experiences gained from a Test Case Application. In Lecture Notes in Computer Science/Application and Theory of Petri Nets 1993, Vol. 691, pp. 542-551, Springer-Verlag, 1993. ISBN 3-540-56863-8

[3] Capellmann, C.; Dibold, H.: The Object-Oriented Petri Net Method for the Specification of IN Services. Proceedings of the International Workshop on Intelligent Networks "Software Methods and Tools for IN Services", pp. 63-76, Lappeenranta/Finland, August 1993

[4] Christensen, S.: Message Sequence Charts. User's Manual. January 1997. Available from http://www.daimi.aau.dk/designCPN/

[5] Dibold, H.: Intelligente Netze — Einführung und Grundlagen. Der Fernmelde-Ingenieur, 44 (1990) 4

[6] Harper, R.; Milner, R.; Tofte, M.: The Semantics of Standard ML, Version 1. Technical Reports ECS-LFCS-87-36, University of Edinburgh, LFCS, Department of Computer Science, University of Edinburgh, The King's Buildings, August 1987

[7] Herzog, U.: Petri Net based Modelling of Interactions between Basic Call State Model and Service Logic in an Intelligent Network. Protocol workshop within the 16th International Conference on Application and Theory of Petri Nets, Torino, June 1995

[8] ITU (CCITT) Recommendation Z.100: SDL, 1992

[9] ITU (CCITT) Recommendation Z.120: MSC, 1992

[10] ITU-T SG XI: Q.1200 Series of Recom. for Intelligent Networks, ITU-T, 1992

[11] ITU-T SG XI: Q.1214, ITU-T, 1992

[12] ITU-T SG XI: Q.1219, ITU-T, 1992

[13] Jensen, K.: Coloured Petri Nets. Basic Concepts, Analysis Methods and Practical Use. Volume 1, Basic Concepts. Monographs in Theoretical Computer Science, Springer-Verlag, 1992

[14] Jensen, K.: A Breif Introduction to Coloured Petri Nets. In: Ed Brinksma (Ed.): Tools and Algorithms for the Construction and Analysis of Systems. Lecture Notes in Compueter Science 1217, Springer-Verlag, 1997

[15] Jensen, K.; et al.: Design/CPN Manual. Meta Software Corporation. Cambridge, USA, 1991. Available from http://www.daimi.aau.dk/designCPN/

[16] Peterson, J.L.: Petri Net Theory and the Modelling of Systems; Prentice Hall Inc., Engelwood Cliffs, N.J. 07632,1981

[17] Rasmussen, J.L.; Singh, M.: Mimic/CPN. A Graphical Simulation Utility for Design/CPN. User's Manual, Version 1.5.
Available from http://www.daimi.aau.dk/designCPN/

[18] Thörner, J.: Intelligent Networks, Artech House Inc., Boston, London, 1994

Formal Methods and Customized Visualization: A Fruitful Symbiosis

Tiziana Margaria[1] and Volker Braun[2]

[1] Universität Passau, Innstr. 33, D-94032 Passau (D), `tiziana@fmi.uni-passau.de`
[2] Universität Dortmund, Baroper Str.301, D-44221 Dortmund (Germany),
`volker.braun@cs.uni-dortmund.de`

Abstract. Formal methods and visualization techniques are central for the realization of *user-centered computing* environments: they are responsible for ensuring correctness and comfort of tool usage in application-level development scenarios. Their synergy is the key to a wide acceptance and improved productivity. We illustrate here on a case study how even elaborate formal methods can be profitably used by non experts, providing that they are fully automatable and supported by powerful visualization aids, and how formal methods are the key to a flexible 'look and feel' of the presentation-layer, which becomes a customizable commodity within a user-specific configuration space. This way, an application development tool becomes really capable to evolve together with the application, the project, and single user's needs.

1 Motivation

A central key to the acceptance of otherwise quite hard-to-understand formal methods is an attractive and friendly graphical user interface, including custom presentation of formulas and results. Surprisingly, the converse, although not appearently obvious, holds as well: formal methods can be used as further means to steer the presentation layer of an environment, in order to grant an *application and user-specific* visualization of models and results. In fact, the translation or manipulation of graphical representations for different user communities sometimes requires the use of formal methods.

In this paper we outline the reach of this potential for synergy via several examples of profitable cooperation. Along a user session with the MOSEL-META-Frame environment we show

- how adequate visualization of analysis and verification models can help even unexperienced users to understand results delivered by a decision procedure for monadic second-order logic (in Sect. 2 to 4), and subsequently
- how a formal-methods based *coordination* of heterogeneous analysis and verification techniques can help improving the presentation of the same results by means of custom views (shown in Sect. 5 and 6). Views tailor the graphical presentation for particular users, thus offering on-demand aspect-driven personalization of the presentation layer.

Both aspects are central requirements for the realization of *user-centered computing* environments: users should not be forced to adapt to fixed development formalisms, but rather the environment should support a variety of application- and user specific needs. In a good environment, the available methods must be accessible and applicable independently of the presentation-layer's 'look and feel', which becomes a customizable commodity within a user-specific configuration space. Moreover, this space must be capable to evolve together with the application, the project, and the single user's needs.

The benefits of the formal methods/visualization synergy for user-centered computing are illustrated in this paper by means of its impact on the applicability of a decision procedure for monadic second order logic on strings. Implemented in the MOSEL toolset [8, 9], it can be used for the automatic analysis, synthesis and verification of parametric hardware circuits. The logic is beneficial in practice to solve a class of frequently occurring problems, but technically unknown to the vast majority of the potential users (here, hardware designers). As such, it is a good representative of the methods for which the potential benefit from this synergy is of vital importance: without adequate and flexible visualization aids which allow a presentation layer familiar to the users, even fully automatic techniques like this would not be accepted by practitioners.

The Application Scenario. MOSEL is a toolset for the automatic analysis and verification of MOnadic SEcond order Logic. Its flexible set of decision procedures for several theories of the logic is complemented by a variety of support components for input format translations, visualization, and interfaces to other logics and tools. From a toolbuilders' perspective, its distinguishing features are a *modular design*, tailored for efficiency and expressiveness, and *integration* in META-Frame [17], our heterogeneous analysis and verification environment which in particular provides the visualization capabilities. In the following we concentrate on the interplay between their analysis components and the visualization features.

2 Visualization as a Key to Formal Methods: Work with Models, rather than Formulas

Semantic decision procedures and algorithms, like e.g. the meanwhile widely used model checkers, are inherently independent of their input syntax. This is a central property for visualization purposes: the same information contained e.g. in rather cryptic logic formulas can be cast into an attractive graphical form familiar to the application designer, without undermining correctness or precision of the underlying formal methods. This independence is exploited in MOSEL in order to reach a completely logic-free application layer.

2.1 From Formulas to Automata

MOSEL is centered around a semantic decision procedure for Monadic Second Order Logic over Strings. Particular attention to the kind of underlying models is paid in the definition of its formal semantics, which is interpreted on strings of arbitrary, but *finite*[1] length. The semantics of its formulas is representable

---

[1] An equivalent intepretation for the logic considers arbitrary finite subsets of infinite strings (WS1S). This is however less adequate for modelling hardware.

$$T ::= Id \qquad\qquad\qquad \text{Second-Order Terms}$$
$$A ::= T \subseteq T \mid T + 1 = T \qquad \text{Atomic Formulas}$$
$$F ::= A \mid \neg F \mid F \wedge F \mid \exists Id : F \text{ Formulas}$$

Fig. 1. Minimal Syntax for M2L(Str)

in terms of *finite-state automata* (see [8] for details), which are 1) familiar to any computer scientist and electrical engineer, thus a widely accepted representation of models, and 2) adequate for visualization purposes, since they can be displayed attractively as graphs. Particular properties of the automata (accepting states, shortes paths) are then easily visualizable on the graphs by means of colours, icons etc.. This way, MOSEL's logic needs not being presented to the users anymore, since we may directly convey the same information using equivalent minimal automata.

The concrete syntax of the MOSEL minimal logic is shown in Fig. 1. Formulas F are constructed from atomic formulas A by means of negation $\neg$, conjunction $\wedge$ and existential quantification $\exists$. There are only two types of atomic formulas: the binary inclusion $\subseteq$ and the successor relation $+1$, a 'shift' operator on strings.[2] They can be applied to second-order terms T, which in the minimal logic can only be variables Id denoting sets. All the other constructs of the logic, including in particular first, second-order, and Boolean terms, quantification over first-order and Boolean variables, and derived logical connectives, are definable in this minimal logic.

The formal semantics is thus based on the two finite state automata of Fig. 2, one for each atomic formula. MOSEL's automata are deterministic and complete. Edge labels represent Boolean functions characterizing the transitions, which are implemented via BDD techniques [4], for efficient realization of conjunction and existential quantification.

2.2 Flexible Automata Visualization

Broadly used in application areas as wide as programming languages, distributed systems, and hardware design, finite automata are also among the formats with the best tool support. In particular they are adequate for use and visualization within application-specific tools. This aspect was a central point in designing MOSEL. Due to its modularity, MOSEL is independent of any particular display and visualization tools and formats. This holds in general for the whole METAFrame environment: its graph component, based on the PLGraph library presented in [2], was in fact developed as polymorphically as possible in order to offer a wide range of application-specific look-and-feel features, useful when addressing particular representation needs.

In applications like hardware design (next Section), where BDDs are the

---

[2] In hardware, this can be used e.g. as a right shift operator on bitvectors.

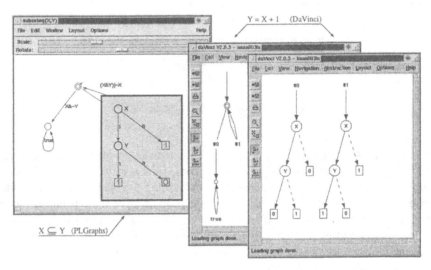

Fig. 2. The Automata for $X \subseteq Y$ and $Y = X + 1$

most widespread data structure, a direct, possibly local, display of the BDD representation for edge labels is such a key feature. As shown in Fig. 2 (left), which displays the 'subset' automaton, clicking on any edge, the BDD representation of its label is *locally* shown in a virtual window.[3] The resulting automata representation is thus *heterogeneous*: the meaning of nodes and edges in the first and in the expanded window differ. As a comparison, Fig. 2 (right) shows how the 'successor' automaton is displayed via the external tool daVinci [5]: two distinct windows represent disjointly information at two different abstraction levels (an automaton, whose edges are labelled by formulas, and a multi-root BDD with a root for each of the formulas). Since these windows are independent, there is no automatic correspondence between the two representations: daVinci does not show which of the root nodes corresponds to which formula, and in fact the BDD root labels #0 and #1 have been added manually.

Working at the semantic level, METAFrame keeps this correspondence even across abstraction levels. A display can then be either local, via a view within the same window as in Fig. 2 (left), or disjoint, by opening an autonomous window. In both cases the correspondences are automatically displayed appropriately.

3 Visualizing Applications: Hardware Circuits instead of Formulas

The attention to adequate coverage of particular *application domains* has to guide the general organization of tools. Concerning MOSEL, the design of its logic reconciles *correctness* aspects (reducible to the correctness of the described minimal logic) with *comfort* requirements.

---

[3] This *window-in-window browser* was introduced in the Intelligent Network services context [18] to show macro expansion within hierarchical telephone service models.

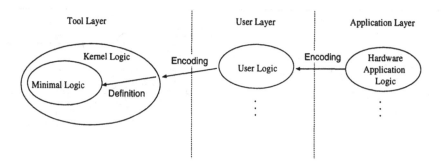

Fig. 3. Layered Logics in MOSEL

A hierarchy of logic layers with increasingly powerful constructs, is related to the minimal logic by either direct embedding or more elaborate encodings as shown in Fig. 3. As explained in [8], the logic itself is successively enriched to offer comfort to the user as well as for direct support of particular application domains (e.g. in our example typical hardware primitives). Due to the principle of implementing outer layers of the logic through successive encodings and definitional extensions to the unique minimal logic, and by making these explicit, the semantic coherence of richer logics with the minimal logic is ensured.

In the following we concentrate on the application domain of verification and synthesis of hardware, dealing with families of parametric sequential circuits [11].

Application-specific Logics: Hardware Libraries. A tool's independence of the internal formalisms is also important at the user input side. In MOSEL we ease the links to established applications-specific notations. For example, the relations shown in Fig. 4 model the behavior of the elementary gates in the hardware application logic. Libraries of logic predicates and correspondig automata for Register-Transfer and gate level libraries are in fact already available.

Compilation from Textual Descriptions. The logic notation is very close to hardware description languages: a direct connection of MOSEL to hardware description languages like VHDL or Verilog could be obtained via an interface to the

```
# circuit constructors as relations
  not(@a,@b) = (~(@a) <=> @b);
  and(@a,@b,@c) = ((@a & @b) <=> @c);
  or(@a,@b,@c) = ((@a | @b) <=> @c);  ...

# D-type flip-flop
    dff(D,Q) = (All t: (t < $) => (t + 1 in Q <=> t in D) & (0 notin Q));
```

Fig. 4. Gate-level Hardware Primitives in MOSEL

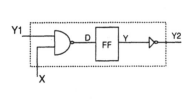
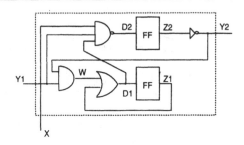

```
rhso1(X,Y1,Y2) =                 rhso2(X,Y1,Y2) =
(Ex D: Ex Y: All t:              (Ex D1: Ex D2: Ex W: Ex Z1: Ex Z2: All t:
nand(t in X,t in Y1,t in D) &    nand3(t in X,t in Y1,t in D1,t in D2) &
dff(D,Y) &                       dff(D2,Z2) &
not(t in Y,t in Y2));            not(t in Z2,t in Y2) &
                                 and(t in Y1,t in Y2,t in W) &
                                 or(t in W,t in Z1,t in D1) &
                                 dff(D1,Z1));
```

Fig. 5. Implementation of the Basic Cells: Circuits and Descriptions

abstract syntax trees delivered by the parsers within the respective compilers, along the lines of [13]. This way it would be possible to use directly the input descriptions delivered by hardware designers.

Compilation from Circuit Layouts. An alternative technique, standard in the hardware community, is the automatic generation of circuit description from symbolic drawings like those in Fig. 5. Symbols appearing there have a direct, one-to-one match with hardware primitives contained in underlying circuit libraries: designers draw circuits in a visual programming style, by selecting, placing, and connecting icons corresponding (in our case) to logic predicates. This *schematic capture* technique is well established, and could be easily supported by MOSEL too, either indirectly (most of the widespread schematic capture tools have an interface to VHDL), or even directly. In fact, it is immediate to derive automatically from drawings like those Fig. 5 the logic descriptions there reported, which refer to the gate-level primitives already introduced. The predicates rhso1 and rhso2 describe the basic cells as a collection of single gates with the appropriate connections. The body of each predicate is a conjunction of calls to predicates of the gate-level library, where the parameter-passing mechanism is used to establish the desired wiring. Internal connections are hidden by means of existential quantification.

In either way, hardware designers unexperienced with monadic second order logic could work exactly as they are used to, and yet profit from the features offered by the MOSEL toolset without any need to master its underlying logic, or even without knowing that the logic exists at all. The next section will show how this is done on a small case study.

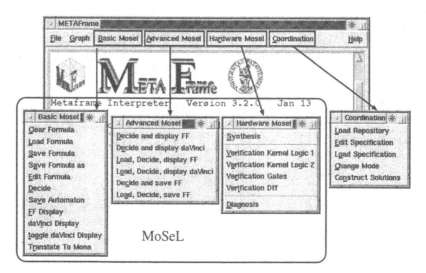

Fig. 6. The MOSEL-METAFrame M2L Environment

4 Visual Hardware Design and Testing

In [12] we conducted a case study from the literature, which illustrates modelling, synthesis, and verification in monadic second-order logic on a class of iterative, parametric, linear systolic arrays. Interesting was that while the two systolic arrays are equivalent, their *basic cells* are not.

Here we revisit the investigation of the basic cells, concentrating on the visualization aspect: indeed, the whole design, analysis, verification, test generation, and debugging process can be carried out in a completely graphical way!

Design via Schematic Capture. A gate-level implementation of the basic cells is shown as schematics in Fig. 5 (top): both rhso1(X,Y1,Y2) and rhso2(X,Y1,Y2) circuits are sequential, have inputs X and Y1, and a single communication output Y2. The corresponding internal logic descriptions used by MOSEL, reported in the same figure, are not shown to the users at all. Still, they match intuitively the structural style chosen by the hardware designers: they are directly readable as netlists of gate-level components.

Evidencing Equivalence. MOSEL's menu-driven GUI allows stating theorems via mouse clicks. From the main window of Fig. 6, in the **Hardware Mosel** modus, the **Verification** menu lets the user select two circuit descriptions to be checked for equivalence. The formula

$$\text{rhso1(X,Y1,Y2)} \iff \text{rhso2(X,Y1,Y2)} \tag{1}$$

requiring rhso1 and rhso2 to possess the same behavior over time, independently of the assignment to the free variables, is automatically generated and checked.[4]

---

[4] This is the typical input/output behavioural equivalence: input X is fed to both circuits and we expect to observe the same sequences Y1, Y2 at the outputs.

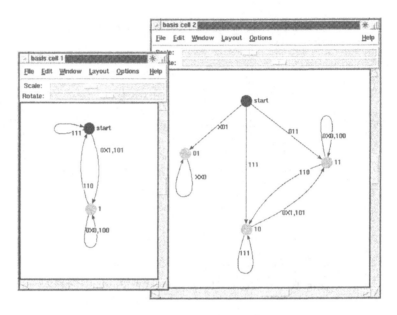

Fig. 7. Synthesized Automata for the Basic Cells

This equivalence statement is refuted. The cause of the failures can be investigated in two ways.

4.1 Visualize the Total Semantics

With the Synthesis command of the Hardware Mosel menu, designers can obtain the automaton corresponding to a selected circuit.

In our example, taking the formula rhso1(X,Y1,Y2) which contains the free variables X, Y1, and Y2, MOSEL synthesizes the automaton of Fig. 7(left), corresponding to the first basic cell circuit. The automaton for the second basic cell is displayed on the right. The chosen presentation form is tailored for hardware designers. Here, the start state is explicitly labelled as such and marked in red on the screen, accepting states (like state 1) are green, non-accepting states would be marked in grey. Edge labels are internally represented as BDDs over the free variables, but here displayed on the graph in cube form, and semantically equivalent to the logic formulas on the edges of the automata.

The refutation of the equivalence statement is clearly correct, since the minimal deterministic automata of the two circuits (Fig. 7), which are known to be unique up to equivalence, differ!

However, for large circuits or less evident differences (e.g. hidden in the edge labels), this form of display may be insufficient for a quick fault detection.

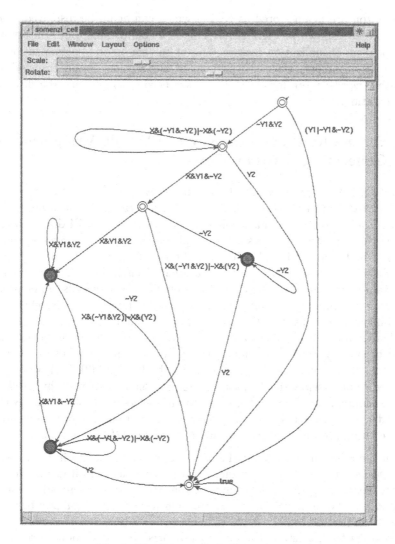

Fig. 8. The Diagnostic Information

4.2 Distinguishing Views

Equivalent diagnostic information can be given as a compact *distinguishing view*, corresponding to an automaton that characterizes only those input sequences which allow observers to distinguish the behaviours of the two circuits.

The distinguishing automaton of Fig. 8 was automatically generated by our attempt of proving formula (1). It does not contain the full semantics of the circuits: here, non-accepting nodes (failure nodes, dark simple circles) are those where the languages accepted by the two circuits differ. Distinguishing views are visualizations for specific purposes: here we address fault detection and also generation of corresponding test patterns. The edge labels in the distinguishing automaton are again stored as BDDs, which can again be displayed along the lines already shown in Fig.2 (left). This is a comfortable, but still application independent visualization feature. To get further in the customization of the visual environment, we need the help of abstraction, graph transformations, in other words: more formal methods are needed to enhance the flexibility of presentation.

5 Formal Methods as a Key to Flexible Visualization: Generating Custom Views

Graphs are not a pure visualization commodity, but offer also a powerful basis for intertool data exchange, crucial for the cooperation between heterogeneous tools. The bidirectional link of MOSEL's automata to the PLGraph library is a central feature, since it allows reading and generating automata which may stem from or be fed into other algorithms and tools of the METAFrame environment.

METAFrame is in fact also a *tool coordination* environment: as shown in other application domains [16, 18, 14], the functionality of single tools can be profitably reused by enabling its combination with other (compatible) tools, whereby compatibility is primarily a matter of interfacing capabilities.

For example, being able to input/output graphs, MOSEL can be additionally used to check imported automata wrt. properties in the M2L(Str) logic. This way the whole environment gains a model checker for M2L(Str), which can be applied to any (semantically suitable) automaton available within METAFrame.

For comparison, Mona [7], meanwhile the 'reference' tool for M2L(Str), does not offer this possiblity due to its I/O format restrictions: not providing interfaces at the automaton level, it can be only applied to M2L(Str) formulas.

In the following we exploit this tool coordination facility in a new way: *for pure visualization purposes*. In fact we will use other algorithms of the META-Frame environment for generating a custom view on the diagnostic information of Fig. 8 to evidence solely the information needed for automatic test pattern generation purposes.

In order to understand the example, we need first to introduce briefly the integration and coordination mechanisms offered by METAFrame. Itself supported by visualization mechanisms, METAFrame's coordination platform uses formal methods in order to construct and check combinations of compatible tools.

5.1 Coordination in METAFrame: Visual Coarse Grain Programming

The METAFrame open tool coordination platform is designed for the combination and coordination of heterogeneous tools: complex tool combinations can be (semi-) automatically or interactively constructed and tested. The specification and on-line construction of combinations are largely graphical, and constitute on-line visual 'meta-programming' at the tool granularity level. This strongly supported coordination allows even non-experts to profitably combine heterogeneous tools to solve own application-specific tasks. In fact, tool coordination

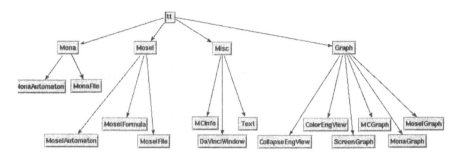

Fig. 9. The Type Taxonomy

becomes a visual programming task, where input and feedback may happen graphically: the presentation of the coordination possibilities is provided in graph form, and during execution of the tools, the progressing of coordination can be followed on the coordination graph.

The underlying internal formalisms and machinery are formal methods (a linear time logic [14] with corresponding automatic synthesis algorithms [15]), which can be completely hidden to the user. As will be shown on the example, even complex coordination tasks can be specified and carried out in an intuitive and largely visual manner. Interactive experimenting with tool combinations is supported by METAFrame's hypertext system (for browsing documentation) and by the automatic tool synthesis mechanism that generates coordination code from loose graphs, interpreted as application-specific specifications. Exactly these features are also used for the loose tool coordination within ETI, STTT's online Electronic Tool Integration service [15].

We are now going to illustrate the use of all these features on our example, leaving the more detailed treatment of the formal background to the cited works.

5.2 Visualizing the Tool Coordination Repository

Tools and algorithms available in METAFrame are kept in a *repository* which forms a collection of basic reusable components considered atomic on the coordination level. Components include elementary tools, but also transformations to bridge the gap between different representations or I/O formats.

Specifications of coordination tasks express constraints on admissible tool combinations. They may be input in completely graphical form (e.g. the loose graph of Fig. 12), which is automatically translated into a *temporal logic*, SLTL [16].

Coordination tasks themselves are automatically synthesized from SLTL specifications, constitute visual programs presented as graphs, and are directly executable. The corresponding high-level programming language is specifically designed for the combination of complex component programs, which may be heterogeneous, e.g. written in different programming languages.

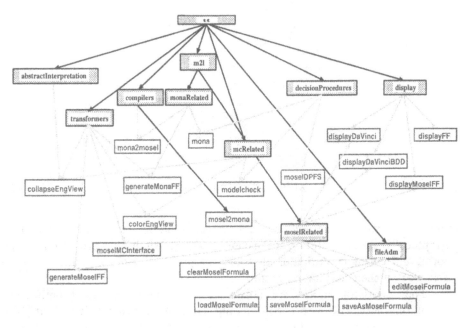

Fig. 10. The Activity Taxonomy

Data and Functional Views: the Taxonomies. The functionality and compatibility information among different tools and formats is captured in two taxonomies, which express properties of data formats (**types**) and of available components (**activities**) respectively. In our setting, concerning a formal methods excerpt of the METAFrame repository, the relevant taxonomies are shown in Fig. 9 and 10, displayed by the hypertext system resp. in our abstract modelling. For example, among the abstract data formats used in the MOSEL toolset are automata (**MoselAutomaton**) and formulas (**MoselFormula**), and among the MOSEL functionalities (**moselRelated** group) there is the mosel2mona activity, which is also a compiler (**compilers** group).

The Formal Methods Coordination Universum. The taxonomies offer visual representations of partial views on the universum of formal methods. Concrete tools and algorithms are abstractly seen as type transformers whose input types range over predicates of the type taxonomy and whose functionalities range over predicates of the activity taxonomy. For example, the **modelcheck** activity takes a **MCGraph** and returns a **MCInfo**.

This way, the (automatically generated) Coordination Universe shown in Fig. 11 (left) models all the possible coordination sequences supported by a given repository, in our case filled with some of the formal methods tools of the METAFrame environment: those based on M2L(Str)(MOSEL and **Mona**), abstract interpretations, and model checking.

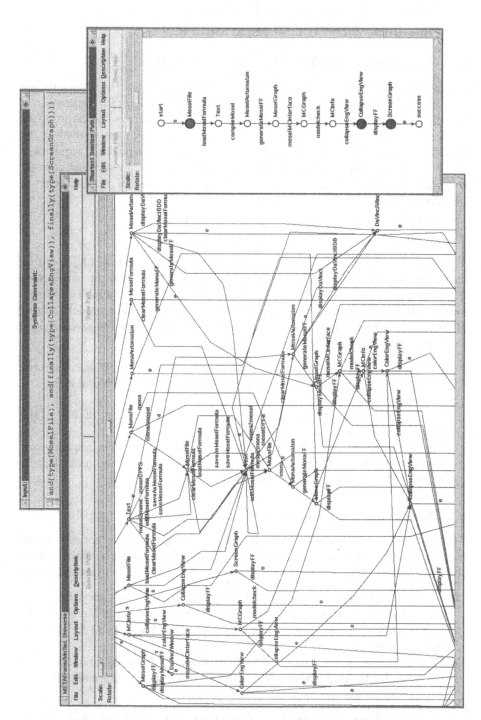

Fig. 11. The Universe, and the Synthesized Coordination Solution

6 Visual Coordination in Practice: Obtaining the Test Generation View

6.1 Defining a Custom View for Test Generation

For test generation purposes, already the small automaton of the distinguishing view (Fig. 8) contains inessential or redundant information. To generate test sequences, hardware designers just need to know the (shortest) paths from the start state to *any* non-accepting state: if the two circuits were equivalent, all the states would be accepting. Moreover, since any distinguishing sequence is a valid and sufficient test, it is sensible to concentrate on the shortest ones for test performance purposes. The optimal diagnostic information is therefore much more compact:

1. all the non-accepting states collapse to a single one,
2. any outgoing path from that state is deleted.

This particular example of graph transformation is quite simple, and could be easily programmed by hand. However, a general method for *property-driven model collapse* based on *abstract interpretation* techniques is available in META-Frame and could be profitably applied for this purpose. Using model collapse offers in fact several advantages wrt. a direct programming of the desired view: the resulting graph transformation tool

- is independent of the particular property, since it addresses a logic (here the full mu-calculus [10]). Thus the view definition process is reusable for any property expressed in the logic,
- can be used without programming, since it is visually specifiable at the coarse-grained tool coordination level,
- can be immediately tested, since the resulting coordination sequence (or sequences, if more than one) are presented as a graph whose paths can be executed by the user (rapid prototyping).

6.2 Specifying Coordination Tasks: A Loose Graph

The specification of the coordination task can be given in an entirely visual manner within the METAFrame Coordination menu. This is made available within the main MOSEL window, as shown in Fig. 6. To define and obtain views we must:

1. Load the tool repository (command Load Repository), which implicitly generates the large universe shown in Fig. 11. This universe is hidden to the user, and shown here only to give a flavour of the background for the synthesis algorithm.
2. Formulate the corresponding loose specification of the coordination task (command Edit Specification) graphically, as in Fig. 12. Informally, the desired specification is:

   Given a MoselFile, *build a* CollapseEngView, *and display it on* screen.

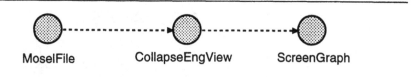

Fig. 12. Specifying the Coordination Task as a Loose Graph

This completely declarative specification does not contain any hint of the name or kind of algorithms which may be used in the solution. A *loose graph* as shown in Fig. 12 represents exactly this information:[5] edges in the graph in fact mean that a transformation in one or more steps must be found, linking the two adjacent nodes. The specification is translated into the textual formulation of Fig 11(top), which is the internal tool-oriented representation.

3. Ask for automatic construction of the solution space (Construct Solutions).

Note that this procedure is parameterized in the concrete property to be used, therefore it defines an entire *class of views*.

6.3 The Coordination Sequence Graph

The resulting coordination sequence graph is computed on the basis of the configuration universe (the large graph in Fig. 11(left)), which users never see.

What users see is only the solution space. In this case, the single (shortest) tool cooordination sequence shown in Fig. 11 (right) needs 7 elementary tool invocations, which are presented in the view as edges: from loadMoselFormula to display. The thick grey nodes correspond to the types explicitly mentioned in the loose specification, while the rest has been automatically generated by the synthesis algorithm. This result is not trivial at all, since e.g. the system 'must know' that the desired model collapse can be obtained as a by-product of model checking!

The coordination sequence graph corresponds to an interpretable program in METAFrame's coordination language. As such it is immediately executable: just click the command Execute Path in the same window.

6.4 Obtaining the Test Generation View for the Basic Cells

During execution, the participating activities to the coordination sequence are launched by the interpreter of the coordination language. The concrete MoselFile to which the sequence has to be applied is chosen interactively via a file selector box. The subsequent tool invocations invoke the MOSEL decision procedure, generate the distinguishing automaton, and transform it in the correct format to be passed to the model checker.

---

[5] In other projects, this input format, used e.g. in [1], proved to have the closest correspondence with the developer's intuition among a set of alternatives.

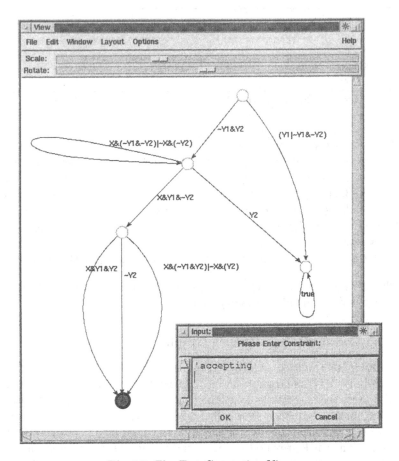

Fig. 13. The Test Generation View

At this point, the property wrt. to which to collapse the model collapse is input in a pop-up editor window, as in Fig. 13 (bottom): in our case, all the non-accepting states must be collapsed together. Pressing the OK button lets the execution progress, resulting in the desired custom view.

Fig. 13 shows the custom view corresponding to the distinguishing automaton of Fig. 8. The model checker has identified all the accepting nodes, and the model collapse algorithm has generated a view where all the non-accepting states have been bundled into a single one which is an end state, i.e. it has no outgoing transitions. This corresponds exactly to our specification of Sect. 6.1, and has been obtained with simple graphical interactions, without any traditional programming.

Related Work. We are not aware of any environment supporting visual, loose coordination among a wealth of formal methods techniques. Closest to our goals are toolsets like ALDÉBARAN [3], EUCALYPTUS [6], or the Concurrency Factory:

they provide collections of tools for the analysis and verification of distributed systems which can be combined on the basis of specific exchange formats. However, their tool coordination capabilities are neither visual nor loose, since tool combinations can only be steered by manual stepwise invocation.

7 Conclusions and Future Work

We have presented an environment for user-centered-computing in which large portions of user interaction, specification, and even complex, coarse-grain programming tasks have been transformed into visual activities. Graphical operations and interactive requirement capture have enabled an advanced use of formal methods even to unskilled users. Central for this achievement is the synergy between formal methods and visualization issues: *strong* [19] formal methods in conjunction with tailored presentation features offer the potential for a powerful and user-friendly application development environment for everybody.

The following questions, posed by one of the referees, are currently still open:

- *The visualisations which the system produces are compelling but I wonder how easy it will be to maintain a clean separation between logic and visualisation were we to extend the ranges of logics and visualisation styles (and allow users to combine these).*
- *It would be interesting to know how much engineering it takes to produce the sort of interaction between logic and diagrams which they demonstrate. Is this simply a specialist task or is it made easy for users in some way (e.g. by ensuring that any form of display will work with any underlying logic)?*

Questions like these characterize the background of the ETI service described in [15], which went online a few weeks ago: the ETI project can be seen in fact as a large-scale trial for the applicability, accessibility, and scalability of the approach proposed here.

Acknowledgement The authors are indebted to Claudia Gsottberger for fruitful discussions and for her valuable help in the realization of this paper, as well as to the reviewers for invaluable observations and comments.

References

1. V. Braun, T. Margaria, B. Steffen, H. Yoo, T. Rychly: *Safe Service Customization*, Proc. IN'97, IEEE Communication Soc. Workshop on Intelligent Network, Colorado Springs, CO (USA), 4-7 May 1997, IEEE Comm. Soc. Press
2. M. von der Beeck, V. Braun, A. Claßen, A. Dannecker, C. Friedrich, D. Koschützki, T. Margaria, F. Schreiber and B. Steffen: *Graphs in* METAFrame: *The Unifying Power of Polymorphism*, Proc. TACAS'97, Enschede (NL), April 1997, LNCS 1217, Springer Verlag, pp. 112-129.

3. Bozga M., Fernandez J.-C., Kerbrat A., Mounier L., *Protocol verification with the ALDEBARAN toolset*, Int. Journ. on Software Tools for Technology Transfer, Vol.1 No.1/2, Springer Verlag, Nov. 1997.

4. R.E. Bryant: *Graph-based algorithms for boolean function manipulation*, IEEE Trans. Computing, vol. C-35(8), August 1986, pp. 677-691.

5. daVinci is available at ftp://ftp.uni-bremen.de/pub/graphics/daVinci

6. H. Garavel: *An Overview of the Eucalyptus Toolbox*. Proc. of the COST 247 Int. Workshop on Applied Formal Methods in System Design, University of Maribor (SLO), June 1996, pp. 76-88.

7. J. Henriksen, J. Jensen, M. Jørgensen N. Klarlund, R. Paige, T. Rauhe, A. Sandholm: *Mona: Monadic second-order logic in practice*, Proc. of TACAS'95, Århus (DK), May 1995, LNCS 1019, Springer Verlag, pp. 89-110.

8. P. Kelb, T. Margaria, M. Mendler, C. Gsottberger: MOSEL: *A Flexible Toolset for Monadic Second-Order Logic,"* TACAS'97, Enschede (NL), April 1997, LNCS 1217, Springer Verlag, pp. 183-202.

9. P. Kelb, T. Margaria, M. Mendler, C. Gsottberger: MOSEL: *A sound and efficient tool for M2L(Str),"* Proc. CAV'97, Haifa (Israel), Juli 1997, LNCS, Springer Verlag.

10. D. Kozen: *Results on the Propositional μ-Calculus*, Theoretical Computer Science, Vol. 27, 1983, pp. 333-354.

11. T. Margaria: *Fully Automatic Verification and Error Detection for Parameterized Iterative Sequential Circuits*, Proc. TACAS'96, Passau (D), March 1996, LNCS N. 1055, Springer Verlag, pp. 258-277.

12. T. Margaria: *Verification of Systolic Arrays in M2L(Str)*, Techn. Rep. MIP-9613, Universität Passau, Juli 1996.

13. T. Margaria, M. Griva, R. Tesio: *Semantic Extraction for the Automatic Verification of VHDL Descriptions*, Techn. Rep. MIP 94-11, Univ. of Passau (D), October 1994, 21 pp.

14. T. Margaria, B. Steffen: *Backtracking-free Design Planning by Automatic Synthesis in* METAFrame Proc. FASE'98, Lisbon (P), April 1998, LNCS, Springer Verlag.

15. B. Steffen, T. Margaria, V. Braun: *The Electronic Tool Integration Platform: Concepts and Design*, Int. Journ. on Software Tools for Technology Transfer, Vol.1 No.1/2, Springer Verlag, Nov. 1997. See also http://eti.cs.uni-dortmund.de.

16. B. Steffen, T. Margaria, A. Claßen: *Heterogeneous Analysis and Verification for Distributed Systems*, In "SOFTWARE: Concepts and Tools", vol. 17, N.1, pp. 13-25, Springer Verlag, 1996.

17. B. Steffen, T. Margaria, A. Claßen, V. Braun: *The* METAFrame'95 *Environment*, Proc. CAV'96, Int. Conf. on Computer-Aided Design, Juli-Aug. 1996, New Brunswick, NJ, USA, LNCS, pp. 450-453, Springer Verlag.

18. B. Steffen, T. Margaria, A. Claßen, V. Braun, M. Reitenspieß: *An Environment for the Creation of Intelligent Network Services*, in "Intelligent Networks: IN/AIN Technologies, Operations, Services, and Applications – A Comprehensive Report" Int. Engineering Consortium, Chicago IL, 1996, pp. 287-300 – reprinted in *Annual Review of Communications*, IEC, 1996, pp. 919-935.

19. P. Wolper: *Where is the algorithmic support?* Strategic Directions in Computing Research, Concurrency Work. Group, Pos. Statement, ACM Computing Surveys, 28A(4), Dec. 1996,http://www.montefiore.ulg.ac.be/ pw/sdcr/concurrency.html

Using a Visual Formalism for Design Verification in Industrial Environments

Rainer Schlör[1], Bernhard Josko[1] and Dieter Werth[2] *

[1] OFFIS , Dept. Embedded Systems
Escherweg 2, D–26121 Oldenburg, Germany
{Schloer,Josko}@OFFIS.uni-oldenburg.de
[2] Siemens AG, AUT GT 23
Gleiwitzer Str. 555, D–90475 Nürnberg-Moorenbrunn
Dieter.Werth@nbgm.siemens.de

Abstract. This paper reports experiences and results gained during the evaluation of the visual formalism STD as specification method for formal verification, performed in cooperation with industrial partners. The visual formalism STD (Symbolic Timing Diagrams) was developed continuously since 1993 by OFFIS as a specification method, which satisfies several needs: (1) It is based on the principles used in the familiar notation of timing diagrams (as conventionally used by hardware designers). (2) It is a method amenable to formal verification, using state-of-the art verification tools efficiently (in particular, symbolic model-checking). (3) It supports *compositional* verification, which is an approach to verify large designs in a compositional way (breaking up proofs of requirements stated for a full design into a sequence of smaller proof tasks, which imply the global proof task). The formalism (with the supporting tools) has been integrated into an established verification environment (CheckOff–M), which allows to verify industrial–scale designs by model–checking.

1 Introduction

Formal verification is – in terms of the industrial awareness – a rather new method. The breakthrough came with the invention of *symbolic model checking*, and the availability of systems supporting this new verification technique ([5, 3]). The idea of symbolic model checking is to avoid explicit representation of the states of a finite state machine by representing its transition relation symbolically using OBDD's ([2, 1]). Using this method, large ASIC–designs (in the order of up to 300 state bits, depending on the code structure) can today be handled to perform verification algorithms such as reachability analysis or – more generally – model checking.

Model–checking takes a *model* (finite state machine) derived from a high-level description (e.g. VHDL–description) and a *property* (requirement) stated in a logic language (e.g. μ–calculus, CTL, or linear temporal logic). Given a

* This work was supported by the ESPRIT project No. 23037 (FORSITE).

particular property, it answers the question, if **all possible model behaviours** satisfy the given property. In this sense, the method is **complete** (or exhaustive, or providing 100% coverage), by contrast to the increasingly incomplete treatment obtained by conventional simulation.

The convincing cases of model checking are those, where by use of the method an error could be found, which would be hard or even impossible to find by simulation. The literature commonly uses such examples as arguments for the usefulness of the method. The converse case is harder to measure: Assume that a certain set of properties has been verified (proven *correct*), what degree of safety has been reached? It may be that some critical property has been overlooked, and it may also be the case that properties have been incorrectly specified.

In particular, the direct use of logic languages for the specification of properties is error prone and hence not acceptable for use by normal hardware designers. Languages like μ–calculus are difficult to understand, and even temporal logic is only adequate for specialists. The current solution is to define (application specific) *property specification languages* (with intentionally restricted power), which give the 'average' designer the expressive power which is needed for typical specification tasks at a minimum risk of constructing false specifications.

The verification environment CheckOff–M [4] (originally developed by Siemens, where it is called CVE) is provided with a textual language for property specification called CIL (CVE Interval Language). CIL uses intuitive operators like **within** and **during** to express temporal relationships between state assertions (which are VHDL–boolean expressions). By contrast, the language STD (Symbolic Timing Diagrams) relies on the intuitiveness of graphical patterns to denote temporal relations. Our experience is that this idea is attractive in the view of industrial designers. In the course of the FORSITE project, STD has been integrated into CheckOff–M, so that properties can now be specified either in the sentential formalism CIL or in the graphical formalism STD (whatever appears to be more convenient).

There are, to the best of our knowledge, currently three main approaches by other research groups to use visual formalisms in the context of formal verification. The work reported in [16] describes a visual logic called RTFIL (real–time future interval logic), which is a visual logic with formulae that resemble timing diagrams. A very important property of this logic (and the supporting tool set) is that it is a real–time logic, which allows *visual reasoning*. Another extensively studied class of visual specifications is reported in [12]. One strong aspect in this work is the strive to make timing diagrams more expressive, for instance by allowing variables (with quantification) in timing annotations. This makes the formalism much more expressive (at the price of higher verification complexity when this feature is used). A third approach is reported in [8], where a class of visual specifications called *constraint diagrams* is defined, which is semantically based on the duration calculus. As yet, the main direction of the work is to support a methodology to derive correct systems by specification transformation.

Overview. The paper continues with the following parts: Section 2 introduces the language STD and illustrates the current tool support for STD. Section 3 describes the application which has been used for evaluation of STD and summarizes the verification results obtained so far. The main point in section 4 is that in general we have to work with *compositional verification*, i.e. we must verify components of a large design in isolation (due to complexity reasons). Section 5 contains conclusions drawn from the evaluation project and points out future improvements and research goals.

2 STD: specifications involving independent events

2.1 Syntax of STD

Consider the following stability property, expressed in linear temporal logic (LTL):

$$\textbf{always}(P \rightarrow \textbf{next}(P \textbf{ unless } Q))$$

which means in natural language: "whenever a system state occurs which satisfies assertion P, then all following states (starting from the next state after the current state) are required to satisfy assertion P, unless a state occurs which satisfies assertion Q (which might never happen)." The state which satisfies Q itself does not need to satisfy P (a subtle point to be noted, since it is important for verification).

Figure 1 shows the statement of this property in the concrete syntax of STD as supported by the tools. (Note that the "Units" line is not relevant for the interpretation of STD).

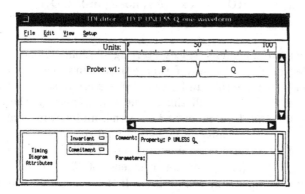

Figure 1. Example of stability property in STD

This is an example of a *linear* STD diagram, i.e. a diagram which has only one waveform, denoting a (symbolic) time line. The following statements apply to linear STD diagrams:

- the time line is depicted as a sequence of regions. The point where two regions are separated, is called an event. The event condition (trigger condition) is drawn flush left within the region which follows the event
- a time line (called a *trace*) has a name (here 'w1'), which may be an arbitrary name to distinguish the trace (so called *probe*-trace), or the name of a port, whose values are specified
- the point in time t, where the diagram is activated, corresponds to the left borderline of the waveform area.

In general, an STD diagram has several time lines (traces). For illustration, we show a formulation of a similar property using two probe-traces in figure 2.

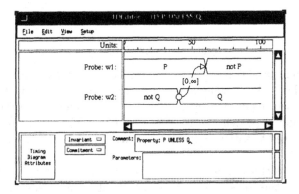

Figure 2. Example of stability property in STD. Equivalent formulation in LTL: **always$((P \wedge \text{notQ}) \rightarrow \text{next}((P \wedge \text{notQ}) \text{ unless } Q))$**

Here we have two timeliness (traces) starting in a common moment t (denoted again by the left border of the waveform area). The first one (trace 'w1') states, that from moment t on, either all following states are P–states, or there occurs a change to a **not** P state (call that event $e2$). Similarly, the second trace (trace 'w2') states, that from moment t on, either all following state are **not** Q–states, or there occurs a change to a Q state (call that event $e1$). What makes the semantics of the diagram meaningful is the so–called precedence–constraint from $e1$ to $e2$, which requires that the occurrence of $e2$ is only allowed, if $e1$ has occurred previously (or at the same time). If this constraint were not present, then the occurrence times of $e1$ and $e2$ would be unrelated (all combinations $e1 < e2$, $e1 \parallel e2$, $e1 > e2$ would be allowed) and the semantics of the diagram would be void.

The constraint used in figure 2 has as timing annotation the interval $[0, \infty]$, which states that the difference between the occurrence time of $e2$ and the occurrence time of $e1$ must be in the interval $[0, \infty]$, i.e. ≥ 0. In general, it is allowed to use any interval with natural number bounds as constraint annotation:

$[m, n]$, $[m, \infty]$, where $0 \leq m \leq n$, are further legal interval specifications for constraints.

2.2 Representing interleaving in STD

A general discussion relates to the interleaving (or partial order) interpretation assumed for events on different traces within a STD diagram. This is unusual compared to the standard case of simulation traces, where same horizontal position of events means simultaneous occurrence. Originally, this semantics was motivated by the observation, that in a distributed system there exists no notion of perfect synchronicity. Communication has to be established then by handshake (request/acknowledge handshake protocol).

However, even in a synchronous system we sometimes want to specify that the relative occurrence times between certain events are irrelevant. Consider the following specification example:

"A circuit has a boolean valued input x_in and a boolean valued output x_out. It is required that (strictly) between two rising edges of x_in there is exactly one point of time, where x_out is high."

There are several cases of a correct implementation:

```
 1    a)
 2         x_in   ___|-------|_____|...
 3         x_out  ___|-|_____...
 4    b)
 5         x_in   ___|-------|_____|...
 6         x_out  _____|-|_____...
 7    c)
 8         x_in   ___|-------|_____|...
 9         x_out  _____|-|_____...
10    d)
11         x_in   ___|-------|_____|...
12         x_out  _____|-|___|...
```

This can be specified in STD by the diagram shown in figure 3.

We can use this diagram to explain three further points about STD:

- Constraints may also start from the activation moment t of the diagram; this moment is represented by the upper–left corner of the waveform area.
- Constraints may be strong (so–called *mandatory* constraints, drawn as solid arrows) or weak (so–called *possible* constraints, drawn as stippled or dashed arrows). Violation of a strong constraint means violation of the diagram (by some given system behaviour), while violation of a weak constraint simply 'terminates' the diagram (restrictions expressed in the diagram do not further apply). The diagram uses one weak constraint (with interval annotation $[1,1]$), from the activation moment to the event where x_in changes from '0' to '1'. This means, that whenever the diagram is activated by state x_in = '0' at moment t, then the diagram is 'interested' only in the case that at the next step the value of x_in has changed to '1' (and at the same time, x_out has assumed the value '0'). In this case, the further behaviour must follow the constraints expressed in the diagram; otherwise, the diagram is terminated and 'waits' for a new activation.

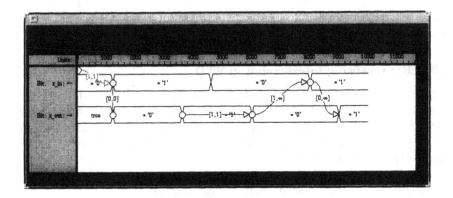

Figure 3. STD–diagram for interleaving specification

- the condition **true** at the beginning of the trace associated with x_out means don't care (since any value of x_out satisfies **true**).

Properties of this kind are difficult to express naturally in a '1–dimensional' formalism. The example shows, how the second dimension can be useful to express timing independence. Note also, that the specification remains equally simple for the following extension of the requirement: "[A circuit has a boolean valued input x_in and a boolean valued output x_out.] It is required that (strictly) between two rising edges of x_in there are exactly **two** points of time, where x_out is high."

2.3 Example of a typical specification pattern in STD

We conclude this introduction and semantical motivation of the formalism STD with an example taken from an actual application (introduced in section 3). It turns out that actual applications often require a certain sequence of events to perform a given function. The following example stems from a particular hardware component verification.

This component (called 'lbrdy') has the following interface (VHDL–entity declaration):

```
1  ENTITY lbrdy IS
2  PORT (
3          adrinc   : IN std_ulogic;    -- pci_address_increment
4          adval    : IN std_ulogic;    -- pci_address/data_valid
5          bover    : IN std_ulogic;    -- burst_over signal
6          clk      : IN std_ulogic;    -- clock signal
7          crdyena  : IN std_ulogic;    -- combinatorial ready_enable
8          crdytyp  : IN std_ulogic;    -- combinatorial ready_type
9          nrdy     : OUT std_ulogic;   -- not_ready
10         rdyena   : BUFFER std_ulogic;   -- ready_enable
11         rdy_ss   : BUFFER std_ulogic;   -- synchronized asynchronous ready
12         ready    : IN std_ulogic;    -- external ready signal
13         reset_n  : IN std_ulogic;    -- lowactive reset signal
14         rtoval   : IN std_ulogic_vector (7 DOWNTO 0); -- not_ready_timeout_value
15         setrto   : OUT std_ulogic;   -- set not_ready_timeout_flipflop
16         start    : IN std_ulogic     -- start burst
17      );
18  END lbrdy;
```

In the handbook (documentation) of the designer there was already an informal timing diagram used to explain one particular function of the component, which led to the formal STD–diagram shown in the next figure.

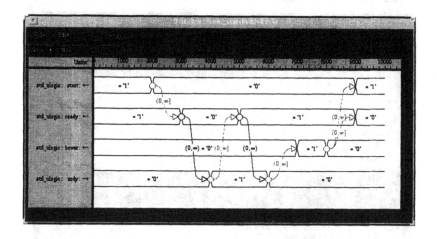

Figure 4. Diagram *com_startReadyBover*

This diagram states after an initial **start** pulse, a negative pulse on signal **ready** is answered by a delayed pulse on signal **nrdy**. The function is concluded by a **bover** (burst over) pulse. There is one new constraint type used in the diagram: A constraint with timing annotation $(0, \infty)$ (more generally, $[m, \infty)$) expresses an *undetermined* bound on the reaction time of the destination event relative to the source constraint of the event: The reaction is required to occur *eventually* in the sense of temporal logic. This type of constraint is used to verify that under all circumstances (environment interactions) the intended component reaction is actually performed. Diagrams of the shown kind are typical. They consist of a sequence (or ordered set) of events, which give the context (or premise) of the requirement, typically adjusted using weak constraints, and the required reactions, usually at the end of the diagram or even woven into the sequence of expected events using strong constraints.

2.4 Management of STD specifications

For the example introduced in the previous subsection, we illustrate how full specifications are handled in the actual verification environment. A tool called *timing diagram manager* (TDM) is used to maintain the actual specification database (sets of diagram declarations) and to allow access to (the creation or modification of) individual diagrams in the set using the timing diagram editor (TDE).

Typically, the verification of a property of the kind shown in figure 4 requires a number of assumptions, stated as diagrams. The next figure shows part of the actual specification developed for the verification.

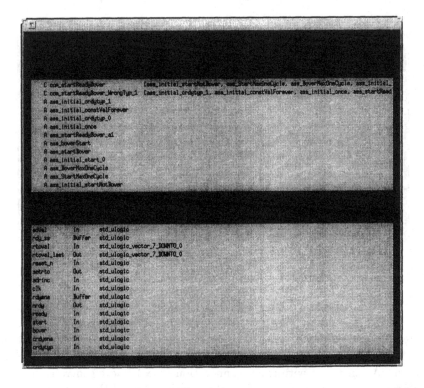

Figure 5. STD specification for *LBRDY* component

A STD *specification* is logically organized into a set of *assumption–diagram declarations* (shown in the manager prefixed with letter A), a set of *commitment–diagram declarations* (shown in the manager prefixed with letter C), and a set of *specification clauses*, where each specification clause R is of the form:

R : **Prove** comm–diagram–name **provided that**
      [assm–diagram–name$_i$ | $i = 1 \ldots k$] **holds.**

This specification clause states, that diagram named comm–diagram–name must be verified about the system (component), under the assumption that the system (or component) environment satisfies the k ($k \geq 0$) assumption diagrams referenced in the clause.

In the current implementation of TDM, the corresponding specification clause is implicitly given with the declaration of the commitment diagram, where the actual set of assumptions used in the specification clause is shown in square brackets behind the name of the commitment. In the actual example shown, it can be seen that a number of assumptions were necessary to find before the commitment shown in figure 4 could be verified.

Typically these assumptions are needed to exclude irregular interferences, e.g. reset, asynchronous change of operation mode, pulse duration requirements, and so on. The task of finding a minimal and sufficient set of assumptions is not trivial, and usually requires several rounds of iteration. Fortunately, each failed verification produces a counter–example, which can be visualized either by conventional VHDL–simulation tools (e.g. Synopsys), or displayed in the TD–editor. This usually gives the clue how an according STD diagram must be formulated which excludes this particular counterexample. We note that in particular at this stage of the verification process, the advantage of using a visual formalism (STD) is very obvious in the view of hardware designers.

2.5   Formal semantics of STD specifications

A full explanation of the formal semantics of STD is beyond the scope of this paper. The interested reader is referred to the publications [6] and [10].

3   Introduction to the PMIO design

In this section, we give a short overview of the design (PMIO) which was used in the verification project, from which the examples used in this paper are taken.

3.1   Overall description of the PMIO design

One of the main products from Siemens Automation is the automation control system SIMATIC. SIMATIC stands for a highly sophisticated family of electronic components needed in the world of industrial automation. Some of these components are conventional CPU's, specialized CPU's, controller and bus interface circuits.

Because inside of the SIMATIC system there is no PCI bus, the PMIO connects the PCI bus to the internal local bus, used by the automation system. The ASIC consists of an PCI– and a local bus interface.

3.2   Block structure of the PMIO design

The block structure of the PMIO design is shown in the next figure.

The various components are listed in the following table.

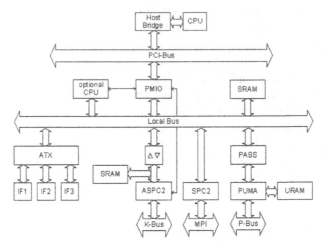

Figure 6. Typical PMIO application

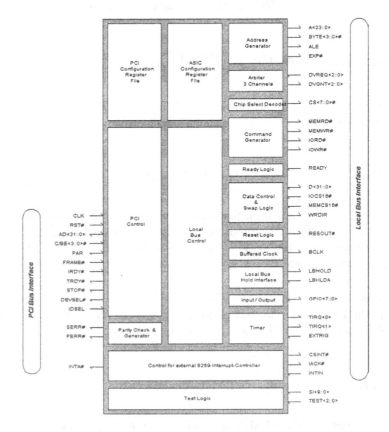

Figure 7. Block structure of PMIO application

Component	comments
PCI Configuration Register File	SpeedChart design
PCI Control	Planned for verification
Parity Check & Generator	SpeedChart design
ASIC Configuration Register File	SpeedChart design
Local Bus Control	SpeedChart design
Address Generator	
Arbiter	Planned for verification
Chip Select Decoder	
Command Generator	
Ready Logic (LBRDY)	Examined by verification (cf. section 2)
Data Control & Swap Logic	
Reset Logic	Examined by verification
Buffered Logic	
Local Bus Hold Interface	Examined by verification
Input/Output	
Timer	Examined by verification
Control for ext. 8259–Interrupt Controller	
Test Logic	

Although no actual errors could be found in the performed examinations (the design was already well tested), in some cases situations where uncovered where the function of a particular component (e.g. the LBRDY component) turned out to be not very 'robust', i.e. the function could easily be corrupted by changes of the (implementation of) component. Such results gained from verification are also valuable, since they give hints for possible (or necessary) quality improvement of the design, avoiding the risk of a design error before it actually happens.

Due to restricted resources within the verification project, only a small number of components could be examined by verification as yet. The main purpose of the evaluation was to demonstrate the feasibility of the approach (verification using symbolic model–checking with STD within CheckOff–M). Some more components are planned to be examined (as indicated in the table). In particular, the PCI–control part of the design is a good candidate, since an STD specification for the PCI bus has already been independently developed and tested [14].

4 Methodology: Model checking + tautology checking

A further advantage of the formalism STD lies in the fact that it provides a methodology for the verification or requirements of a full design, based on a combination of model– and tautology checking (figure 8).

In order to verify a requirement for a full design, the following sequence of steps (referred to as 'system verification') is performed:

1. Identify all subcomponents relevant for the requirement
2. Find sufficiently strong sub–specifications for each of the subcomponents and use tautology checking to show that the requirement is implied by the sub–specifications
3. Verify the sub–specifications for each sub–component using model checking. If necessary, apply system verification again for the verification of (large) sub–components.

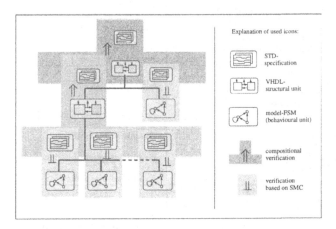

Figure 8. Outline for system verification process

Technically, compositional verification steps are performed using a proof-obligation generator, which converts a VHDL–structural description into a set of logical invariants characterizing the interconnection semantics, then derives (linear) temporal logic characterizations of the involved STD specifications, and constructs the proof obligation, which has the form of an implication in temporal logic. This proof obligation is handled by the tautology checker, which is actually a subcomponent of the model checker which underlies the CheckOff–M system.

4.1 Typical scenarios for system verification

Although system verification was not yet performed on the PMIO design (a good application of system verification was performed recently on a large StateMate design in cooperation with British Aerospace reported in [9]), the example of the verification of the LBRDY shows how system verification will be used to complete the verification. The set of used assumption can be roughly divided into two classes:

1. assumptions which build separate *cases* for verification (e.g., excluding asynchronous reset in a particular verification run using an according assumption)
2. assumptions which must be guaranteed by the behaviour of components in the environment.

The second case induces system verification steps, since these assumptions must be (formally) discharged in order to demonstrate that the particular function will actually be guaranteed in the final design. In other words, the validity of any assumption of the second type must be verified to be a property of the system, implied by the proper cooperations of the design components.

4.2 Guidelines for preparation of designs for system verification

A comprehensive example for system verification using model–checking and tautology–checking with STD is currently under preparation in cooperation with Italtel (ILC design). The design (coded in VHDL) is prepared with a comprehensive documentation, which explains for each functional block of the design:

- Interface definition with explanation of the signals
- Informal functional description
- Requirements

It is important to emphasize, that this style of documentation is not generally standard in industrial practice. For instance, the documentation of the PMIO describes the functional behaviour of the components of the design at different levels of detail, which makes the approach of system verification difficult, since additional communication with the designer becomes necessary. Hence, one vital conclusion from our verification experience is that an appropriate style of documentation should be developed and standardized for designs which are subject to formal verification treatment.

5 Conclusion

The evaluation has demonstrated the usefulness of the method applied to an industrial application (ASIC verification). The experience is that extensive training is necessary in order to make the method and tools accessible to average designers, who are unfamiliar with logic descriptions and the ideas underlying formal verification (in particular, model checking).

The use of a graphical language greatly helps to bridge the gap between the way the designer is thinking (in terms of simulation traces) and actual formal verification. A current research challenge with respect to the specification method is to accumulate typical specification patterns and to provide specification libraries, where useful or common patterns can be easily extracted and instantiated.

With respect to system verification, there is currently only little support for the task of finding appropriate component specifications (which have to be invented from the functional description of the respective component). One line of future research is going to investigate the possibility of an automatic construction of such local specifications.

References

1. R.E. Bryant. Graph-based algorithms for boolean function manipulation. *Transactions on Computers*, C-35:677–691, 1986.
2. R.E. Bryant. Symbolic boolean manipulation with ordered binary decision diagrams. *ACM Computing Surveys*, 24(3):293 – 318, 1992.

3. J.R. Burch, E.M. Clarke, K.L. McMillan, and D.L. Dill. Sequential circuit verification using symbolic model checking. In *Proceedings of the 27th ACM/IEEE Design Automation Conference*, pages 46–51, June 1990.

4. *CheckOff*, user's guide. Abstract Hardware, 1997.

5. E.M. Clarke, E.A. Emerson, and A.P. Sistla. Automatic verification of finite state concurrent systems using temporal logic specifications: A practical approach. In *Proceedings of the 10th ACM Symposium on Principles of Programming Languages*, pages 117–126, 1983.

6. Werner Damm, Bernhard Josko, and Rainer Schlör. Specification and verification of VHDL-based system-level hardware designs. In E. Börger, editor, *Specification and Validation Methods*, pages 331–410. Oxford University Press, 1995.

7. H. Dierks and C. Dietz. Graphical Specification and Reasoning: Case Study "Generalized Railroad Crossing". In J. Fitzgerald, C.B. Jones, and P. Lucas, editors, *FME'97*, volume 1313 of *Lecture Notes in Computer Science*, pages 20–39. Springer Verlag, 1997.

8. C. Dietz. Graphical Formalization of Real-Time Requirements. In B. Jonsson and J. Parrow, editors, *Formal Techniques in Real-Time and Fault-Tolerant Systems*, volume 1135 of *Lecture Notes in Computer Science*, pages 366–385, Uppsala, Sweden, September 1996. Springer Verlag.

9. G. Döhmen, H.J. Holberg, and R. Schlör. Verification of the sms design. Technical report, OFFIS, 1997.

10. Konrad Feyerabend and Bernhard Josko. A visual formalism for real time requirement specifications. In Miquel Bertran and Teodor Rus, editors, *Transformation-Based Reactive Systems Development, Proceedings, 4th International AMAST Workshop on Real-Time Systems and Concurrent and Distributed Software, ARTS'97*, Lecture Notes in Computer Science 1231, pages 156–168. Springer-Verlag, 1997.

11. K. Fisler. A logical formalization of hardware diagrams. Technical report, Computer Science Department, Indiana University, September 1994.

12. K. Fisler. *A Unified Approach to Hardware Verification Through a Heterogenous Logic of Design Diagrams*. Dissertation, Indiana University, 1997.

13. K. Fisler. Containment of regular languages in non-regular timing diagram languages is decidable. In Orna Grumberg, editor, *9th International Conference on Computer Aided Verification*, Lecture Notes in Computer Science 1254, pages 155–166. Springer-Verlag, 1997.

14. Swen Masuhr. Formale Verifikation eines ASIC. Diplomarbeit, Universität Oldenburg, 1995.

15. L.E. Moser, P.M Melliar-Smith, Y.S. Ramakrishna, G. Kutty, and L.K. Dillon. A real–time graphical interval logic toolset. In *8th International Conference on Computer Aided Verification*, pages 446–449, June 1996.

16. L.E. Moser, Y.S. Ramakrishna, G. Kutty, P.M. Melliar-Smith, and L.K. Dillon. A graphical environment for design of concurrent real–time systems. *ACM Transactions on Software Engineering and Methodology*, 6(1):31–79, 1997.

17. R. Schlör and W. Damm. Specification and verification of system-level hardware designs using timing diagrams. In *Proceedings, The European Conference on Design Automation*, pages 518–524, Paris, France, feb. 1993. IEEE Computer Society Press.

Automatic Error Location
for IN Service Definition

Volker Braun[1]   Tiziana Margaria[2]   Bernhard Steffen[1]   Haiseung Yoo[1]

[1] Universität Dortmund, Lehrstuhl für Programmiersysteme, Baroper Str. 301,
D-44221 Dortmund (Germany), *steffen@cs.uni-dortmund.de*

[2] Universität Passau, Innstr. 33, D-94032 Passau (Germany)

Abstract. The paper presents a new unique feature of the IN-METAFrame
Service Definition Environment: the automatic generation of diagnostic
location information as a consequence of detecting an error in the design
phase of a Service Logic. Violations of constraints which express frame
conditions for the design (concerning e.g. implementability, country specific standards, and network specific features) are detected by formal
verification techniques. The subsequent error diagnosis and correction is
now supported by a new kind of abstract views, which not only give hints
on the possible source of trouble, but additionally *automatically locate*
the exact occurrence of the constraint violation in the Service Logic.

1 Motivation

Service Definition Environments for the creation of IN-services are usually based
on classical 'Clipboard-Architecture' environments, where services are graphically constructed, compiled, and successively tested. Two extreme approaches
to error handling characterize the state of the art of marketed SD environments:

- The *avoidance* approach guarantees consistency by construction, but the
  design process is strongly limited in its flexibility to compose Service Independent Building Blocks (SIBs) to new services.
- The *creative* approach allows flexible compositions of services, but there is
  little or no feedback on the correctness of the service under creation during
  the development: the validation is almost entirely located after the design is
  completed. Thus the resulting test phase is lengthy and costly.

Our environment conjoins the desirable features of both approaches: the Service
Logic design phase is as free as in the liberal approach, but it offers enough early
error detection and correction techniques, based on a combination of formal
methods, to enable competent users to avoid an even larger class of errors than
covered by the avoidance approaches.

IN-METAFrame is the first and only industrial tool[1] making advanced formal methods applicable for non-specialists. Its implementation is based on the
METAFrame [13] environment.

---

[1] See [1], as well as the position statements [8, 9] in the same issue.

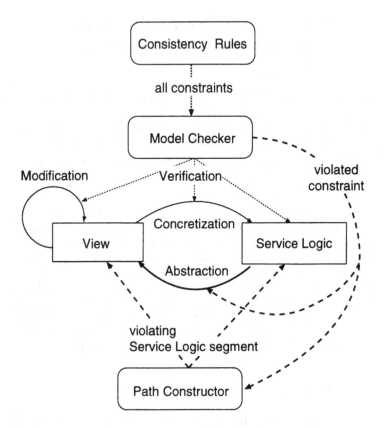

Fig. 1. Error Detection and Location Mechanisms in IN-MₑₜₐFrame

While a detailed description of the tool can be found in [5], in this paper we focus on the interplay of formal verification, abstract views, and property-oriented modal collapse in supporting not only early error detection (as already shown in [5,11,7] and extended in [12] to handle also hierarchically defined services), but also *precise, automatic error location.* This new feature completely eliminates the lengthy, costly searching for the source of the mistake in increasingly large and complex Service Logic graphs, eases the error diagnosis to the point that it becomes evident, and supports timely error correction.

An earlier version of the IN-MₑₜₐFrame SD environment is incorporated in the Advanced Service Design of INXpress, the Siemens solution to Advanced Intelligent Networks, which is commercially available. Presented at various international fairs (e.g. CeBIT'96 and '97), it is already installed at a number of customers (e.g. Deutsche Telekom, South Africa's Vodacom, Finnland's RadioLinja), while further contracts (like Switzerland's PTT and Germany's T-Mobil, Mannesmann, and VIAG) have meanwhile followed. Moreover, INXpress ASD with IN-MₑₜₐFrame is also Winner of the *European Information Technology Prize* 1996, awarded by the European Union through its Esprit-EUROCase organization.

The following section briefly recalls those unique features of our METAFrame based Service Definition Environment which are essential in enabling the implementation of the error location feature. Subsequently, Section 3 shows on a concrete example the currently available error detection facility, and Section 4 illustrates on the same example the improvement of diagnostic capability offered by the new error location feature.

2 Service Definition in IN-METAFrame

The IN-METAFrame environment is constructed for the flexible and reliable, aspect-driven creation of telephone services in a 'divide and conquer' fashion: initial prototypes are successively modified until each component satisfies the current requirements. The entire service creation process is supported by thematic views that focus on particular aspects of the service under consideration. Moreover, the service creation is constantly accompanied by on-line verification of the validity of the required features and the executability conditions for intermediate prototypes: design decisions that conflict with the constraints and consistency conditions of the intended service are immediately detected via model checking (cf. Fig. 1).

The impact of formal methods on service creation is best illustrated by means of the formal verification and abstract views concepts.

- Formal verification allows designers to check for *global consistency* of each design step with implementation-related or service-dependent frame conditions. Being based on model checking techniques [4], it is *fully automatic* and does not require any particular technical knowledge of the user. This simplifies the service design since sources for typical failures are detected immediately.
- Abstract or thematic *views* concentrate on the required global context and hide unnecessary details. They allow the designer to choose a particular aspect of interest, and to develop and investigate the services under that point of view. This supports a much more focussed service development, which concentrates on the design of the aspect currently under investigation. Of particular interest are *error views* that concentrate on the essence of a detected error as explained in Section 3.2.

Both formal verification and abstract views are fully compatible with the macro facility of the environment [12]. This allows developers to introduce *hierarchical structuring* into complex services, e.g. by defining whole subservices or functional units as primitive entities called *macros*, which can be used just as SIBs. As macros may be defined on-line and expanded whenever the interna of a macro become relevant, this supports a truly modular service construction.

The following simple Free-Phone (in the US known as 800-) service will be used for the explanation of the more advanced features of our error detection and location mechanisms, showing their interplay with the formal verification and abstract views.

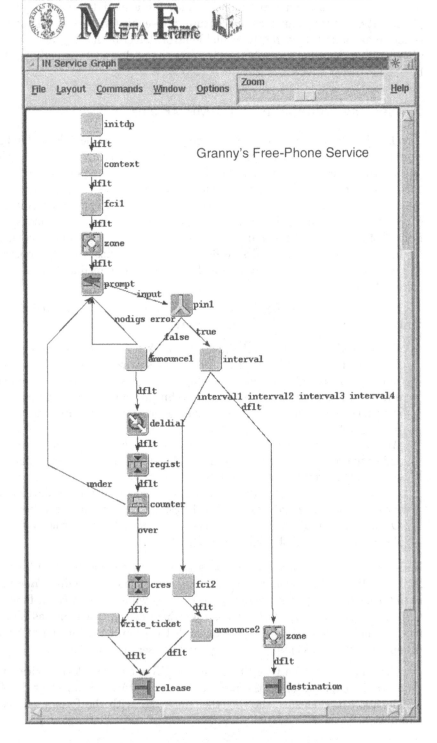

Fig. 2. The Granny's Free-Phone Service.

Granny's Free-Phone The small service shown in Fig. 2 presents the Service Logic graph of a simple kind of Free-Phone (800-service). This particular service is designed for people who would like to incentivate friends and relatives to call them by paying their calls if they are done at a convenient time of the day. In essence, the Service Logic is the following: after a call initialization section common to all services, the caller dials the desired specific Free-Phone number, then a prompt requires entering a Personal Identification Number (PIN) and, if the PIN is correct, depending on the time of the day, the call is either released (in the forbidden time windows) after an announcement, or it is routed to the desired destination number.

3 Early Error Detection

The IN-METAFrame SD environment is unique in providing fully automatic verification of the *global* correctness and consistency of the Service Logic: vital properties concerning the *interplay* between (arbitrarily distant) SIBs of a service can be verified at any time during the design of the Service Logic via model checking in a push-button fashion.

3.1 Formal Verification via Model Checking

A property is global if it does not only involve the immediate neighbourhood of a SIB in the Service Logic graph[2], but also relations between nodes which may be arbitrarily distant and separated by arbitrarily heterogeneous subgraphs. The treatment of global properties is required in order to capture the essence of the expertise of designers about do's and don'ts of service creation, e.g. which SIBs are incompatible, or which can or cannot occur before/after some other SIBs. Such properties are rarely straightforward, sometimes they are documented as exceptions in thick user manuals, but more often they are not documented at all, and have been discovered at a hard price as bugs of previously developed services which obstinately did not want to obey the known rules. They build exactly the kind of precious domain-specific knowledge that expert designers accumulate over the years, which is particularly worthwhile to include in the design environment for future automatic reuse. In the presented environment, such properties are gathered in a Constraint Library, which can be easily updated and which is automatically accessed by the model checker during the verification [7]. Here the model checker of [4] is used, which verifies whether a given model satisfies properties expressed in a modal logic called the modal mu-calculus [2]. In the IN-SD setting:

– the *properties* express correctness or consistency constraints the target IN service is required to respect. They are expressed in the temporal logic SLTL (Semantic Linear Time Logic,[6]). This is a linear-time variant of Kozen's mu-calculus [2], which comes together with efficient verification tools based on model checking;

---

[2] I.e., the set of all the predecessors/successors of a node along all paths in the graph.

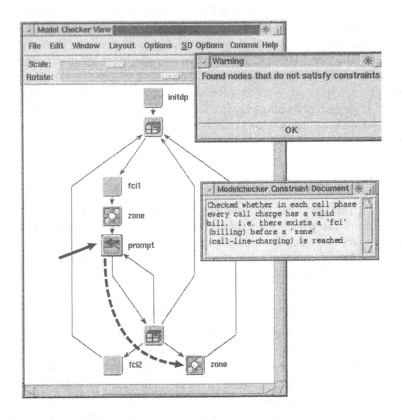

Fig. 3. The Model Checker Finds a Billing Error

- the *models* are the Service Logic graphs, where SIB names correspond to atomic propositions, and branching conditions correspond to action names in the SLTL formulas.

A user-friendly graphical interface, displaying all the relevant data, supports the model checking procedure.

In the example, model checking the service shown in Fig. 2 results in the discovery of erroneous paths in the graph (as shown by the error message window of Fig. 3(top)) which violate some global constraints. The problematic part of the paths starts at the prompt node marked in the figure by the thick arrow, and whose icon is framed in red by the system.

When the model checker detects an inconsistency, a plain text explanation of the violated constraint appears in a window (Fig. 3(middle)). In our example, the violated constraint requests that on each segment (phase) of the call, a bill should be issued before the system can charge for the segment. More concretely, the logic formula describing the constraint is internally expressed as follows:

```
(prompt | initdp) => ( ~zone U (fci | fci1 | fci2))
```

meaning: starting from any prompt or any initdp SIB (which mark the start of a call segment), no zone (billing) SIB should be found without a preceding fci or fci1 or fci2 (calling-line-charging) SIB in the same segment.[3].

To ease the location and correction of the error, till now an abstract *error view* was automatically generated, to evidence only the nodes which are relevant to detect the error.

3.2 Abstract Views and Error Correction

Views are *abstract* service models. As such, they show aspects of actual, concrete models. They are used to hide any aspect of an IN model which is irrelevant wrt. an intended operation. This is useful during the development phase in order to concentrate on specific themes, e.g., the billing or the user-interaction contained in a service, while abstracting from all the rest. This is a solution to the problem of the growing size of the services, which may contain several hundreds of nodes, and which are in their whole unmanageable.

Operationally, views do not differ much from the actual IN models. E.g. they can be loaded and edited in the usual way, however, often with quite dramatic effects: minor modifications on views may correspond to radical structural changes of the underlying concrete model. In addition, views can be *created*, corresponding to the application of an abstraction function, and *applied* to the underlying concrete model, corresponding to the application of a concretization function (cf. Fig. 1). *Execution* of a view means execution of the underlying concrete model.

Error Views. Of particular interest are *error views*: they highlight the essence of a global inconsistency, help locating the error and simplify its diagnosis and correction. Errors can be corrected directly on the error view, and the subsequent view application transmits the modifications to the concrete model.

The design error in our example is now going to be corrected solely with the help of an error view. The comfort of automatic error location offered by our new prototype will then be shown for the same example in the following section.

- The error view of Fig. 3 has been obtained automatically after the failed model checking. The view concentrates on the important nodes for the error diagnosis. All the SIBs which are irrelevant for the detection of this error have been automatically abstracted away and are collapsed in the unlabelled nodes. It is now easy to see that the problem must be located on some path starting at the prompt SIB, which is evidenced by the red framed icon. By inspection of the subgraph starting at that SIB the designer can in fact see that there is a path (indicated by the dashed arrow in the picture) conducing to a zone SIB on which the zone occurrence has no corresponding fci node. The location of the error is eased by the information of the starting SIB and by the description of the violated constraint, but is still done manually.

---

[3] Technically, this is a conditional safety property expressed in the SLTL temporal logic [6], which reads "prompt or initdp implies not zone until fci or fci1 or fci2".

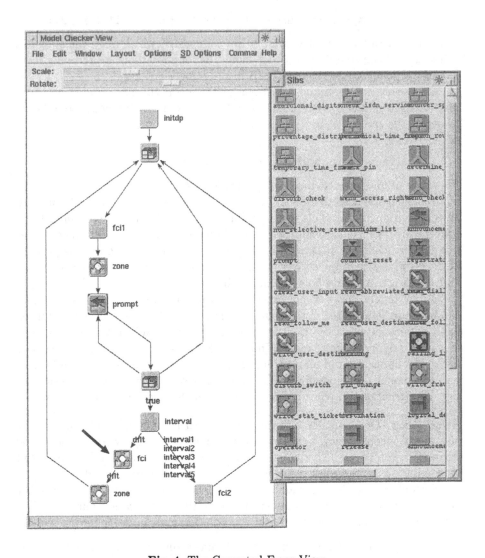

Fig. 4. The Corrected Error View

- Fig. 4 shows the results of the error correction: the view graph has been edited, by inserting the fci node marked by the arrow in appropriate position on the segment.

- The view application command actualizes the corresponding IN-service model (see Fig. 5) which is thus automatically corrected. A new verification via model checking confirms now the correctness of the service.

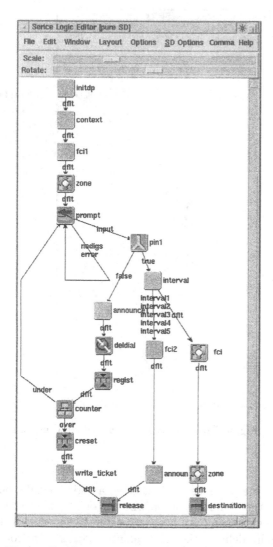

Fig. 5. The Correct Granny's Free-Phone Service

4 Automatic Error Location

Automatic error location improves the previous situation by eliminating the manual search of the erroneous segment of the Service Logic. Since the formulation of properties and correctness conditions is execution-oriented, and expresses observable, traceable run-time behaviours, constraints are path-oriented, i.e. they describe properties or forbid occurrences of SIBs along sequential segments of the Service Logic. Technically this means that constraints can be expressed as linear-time formulas, which is central for our new diagnosis feature.

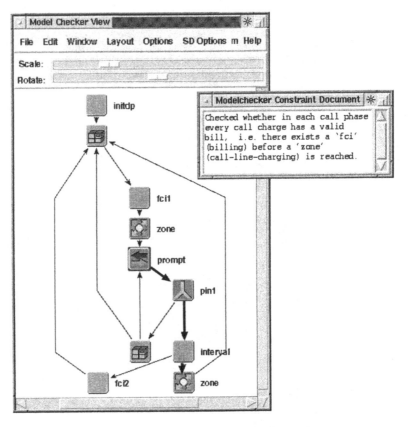

Fig. 6. Error Location for the Billing Constraint

As shown in the previous section, given a logic formula, our model checker does not only return information on whether or not it is generally valid on the whole Service Logic graph (as it is usual for iterative model checking algorithms), but it additionally delivers the exact points in the graph from which violating paths start. It is therefore possible to pass the result of the model checker to an algorithm for the property-driven construction of satisfying paths. More precisely, the negated formula is here used, in order to discover not the correct, but the erroneous paths starting at the given point in the Service Logic graph.

Abstract view presentation is still useful here, but with a different flavour: instead of applying the property-oriented abstraction to the whole service graph, as shown in the previous section, we enhance the diagnostic capabilities by applying the abstraction on the correct portions of the Service Logic, which do not matter here, and leaving the erroneous path unchanged.

This choice offers the best diagnostic support to the designer, since the attention is immediately drawn on the (usually small) erroneous portion of the logic, which is moreover immediately ready for correction, without requiring additional decapsulation operations of abstract nodes as in the previous versions.

Fig. 6 shows the automatically generated error view with error location information for the same service and constraint already presented in the previous Section. The path connecting the (red framed) prompt SIB to the zone SIB over the pin1 and interval SIBs is completely unrolled (it mathches in fact directly the rightmost path of the Granny's Free-Phone Service Logic shown in Fig. 2, which is indeed incorrect, since it violates exactly this constraint!) and it is evidenced by the red colour on the screen. To compare, in Fig. 3 only the starting and terminating point were visible, and the path had to be 'discovered' by the designer via unrolling of the abstract node between them.

Note that this form of automatic error location is the *optimal* support for error diagnosis: the checked properties involve an interplay of several SIBs along the path, which usually describe their relative positioning in a loose fashion. Thus they leave room for several alternatives. As a consequence, the legal error correction is not unique (in the example, one could insert the missing billing SIB at any place on the path between the prompt and the zone SIBs), and a corresponding automatic selection process is not wished, as it would unnecessarily constrain the designer, who usually chooses the appropriate correction according to other (semantic) criteria. In this case, service providers may desire billings to be valid on the longest possible portion of a call segment. Thus the most convenient location for the billing SIB is immediately before the zone SIB.

4.1 Error Location for a Library of Constraints

In reality, a whole library of properties has been defined. Each single constraint is approximately of the size of the property already presented in the previous section, and is easily checked on-line. As shown in Fig. 1, the model checker in fact goes through the list of single properties, and stops with an error message window and with the property-specific generation of an error view as soon as a violation is detected.

To give a feeling of how the mechanism works, in the currently checked library there are, among others, also the following two constraints, concerning different aspects of the service and a PIN check SIB pin1.

Wrong PIN Constraint: Connections can only be established after a successful PIN check.
The following formula

```
pin1 => [false] (~destination U pin1)
```

means: leaving a pin1 SIB along the false branch (i.e. after the input of an incorrect PIN), implies that a destination SIB cannot be reached unless a pin1 SIB is found again. As this second occurrence of a pin1 SIB 'triggers' the constraint again, one can only reach the a destination SIB after leaving a pin1 SIB along the true branch.

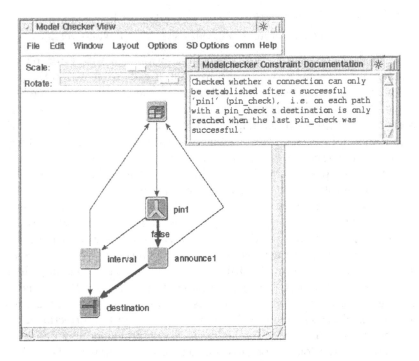

Fig. 7. Error View with Location for the Wrong PIN Constraint

The corresponding error view is shown in Fig. 7 together with the constraint documentation. Being property dependent, the generated view is completely different from the error view of Fig. 6: here, only the **pin1**, **announce1**, and **destination** SIB have remained concrete (and the connecting path is highlighted in red on the screen), and all the other portions of the Service Logic have been abstracted away.

Forbidden Billing Constraint: Unsuccessful calling sections will not be charged. The formula

```
release => (~zone) BU (prompt | initdp)
```

means: going backward from the **release** SIB, which marks unsuccessful call attempts, no **zone** (segment charging) SIB should be met on the segment, which is delimited by a **prompt** or by an **initdp** SIB. The corresponding error view and the constraint documentation are shown in Fig. 8.

Here we see that the logic allows expressing constraints not only along the flow of the call (forward constraints), but also in the opposite direction. Thus, a wide class of causality interactions can be elegantly and concisely formulated, succesfully checked, and efficiently enforced.

Of course this simple example service could have been still easily handled by hand, but current IN services have reached sizes and complexities which demand for automated support for error detection, diagnosis, and correction. Our

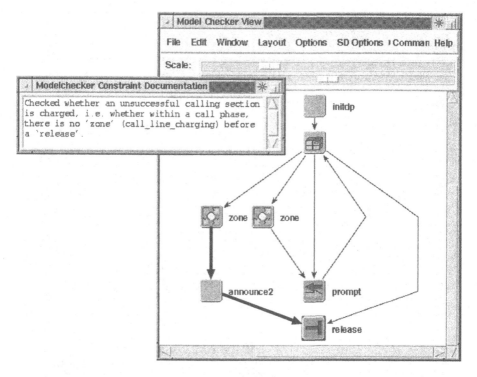

Fig. 8. Error View with Location for the Forbidden Billing Constraint

environment encourages the use of the new method, as it can be introduced *incrementally*: if no formal constraints are defined, the system behaves like standard systems for service creation. However, the more constraints are added, the more reliable are the created services [10].

4.2 Validating and Correcting the UPT Service

To demonstrate the scalability of IN-METAFrame ś error detection and correction facilities, we show how the same techniques impact the treatment of a quite more complex service: the Universal Personal Telecommunication Service (UPT).

The UPT service combines personal mobility with the access to and from telecommunication over a unique number and account. An year ago it was one of the most complex services available. Using a personal identifier, a UPT subscriber can access telecommunication services at any terminal and use those services provided by the network which are defined in the own service profile. Personal mobility involves identifying the location of the terminal currently associated with the subscriber. Incoming UPT calls must be routed to the current destination address, and the associated charge may be split between the calling

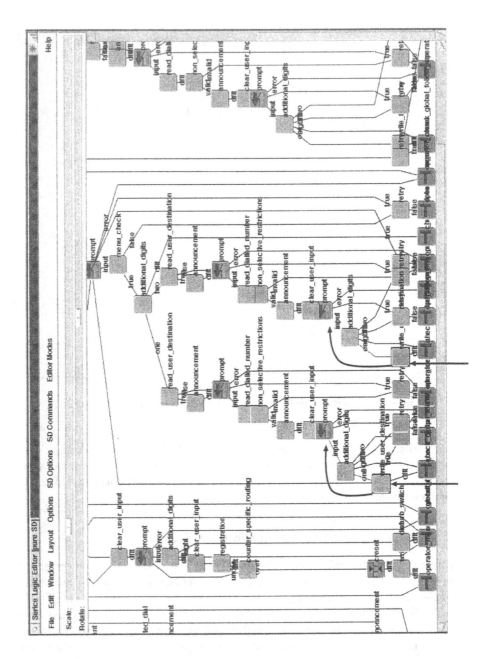

235

Fig. 9. The UPT Service with Two Violations

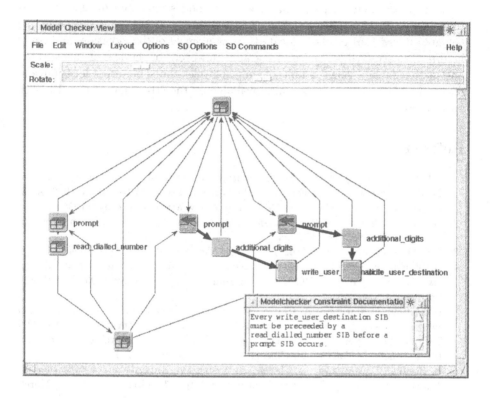

Fig. 10. The UPT Error View

line and the UPT subscriber. Subscribers can use any terminal in the network for outgoing UPT calls, which are charged to their accounts. This requires user identification and authentication on a per-call basis. The use of the optional follow-on feature allows one authentication procedure to continue to be valid for subsequent calls or procedures. The service package can be tailored to the subscriber's requirements selecting from a comprehensive service feature portfolio.

The Service Logic graph, a portion of which is shown in Fig. 9, has 158 nodes and 239 edges: although the manual investigation of complex interdependencies among its SIBs requires a fair amount of effort from a team of designers, its automatic verification by model checking wrt. the constraint library is still done on-line within seconds, even for hundreds of constraints!

For the example constraint

```
write_user_destination => (~prompt BU read_dialled_number)
```

meaning that every write_user_destination SIB must be preceded by a read_dialled_number SIB before a prompt SIB occurs, the automatic error location facility showed that 2 segments of the current Service Logic violate the constraint.

In fact, the marked paths starting at the nodes indicated by the thick arrows in Fig. 9 (the nodes are framed in red on the screen, and the paths are coloured in red too) precisely locate two occurrences of the violation, and could be directly corrected on the service graph. However, for services of this size the abstraction facility plays a quite significant role: the same error location and diagnosis information is in fact also available on the error view of Fig. 10, with only 10 nodes!

References

1. Ed Clarke, J. Wing: *ACM Worksh. on Strategic Directions in Computing Research*, Position Statement of the Formal Methods group, Boston (USA), June 1996. *ACM Computing Surv.* 28A(4), Dec. 1996, http://www.acm.org/surveys/1996.

2. D. Kozen: *"Results on the Propositional μ-Calculus"*, Theoretical Computer Science, Vol. 27, 1983, pp. 333-354.

3. J.K. Ousterhout: *"Tcl and the Tk Toolkit,"* Addison–Wesley, April 1994.

4. B. Steffen, A. Claßen, M. Klein, J. Knoop. T. Margaria: *"The Fixpoint Analysis Machine"*, Proc. CONCUR'95, Pittsburgh (USA), August 1995, LNCS 962, Springer Verlag.

5. B. Steffen, T. Margaria, A. Claßen, V. Braun, M. Reitenspieß: *"An Environment for the Creation of Intelligent Network Services"*, in "Intelligent Networks: IN/AIN Technologies, Operations, Services, and Applications – A Comprehensive Report", Int. Engineering Consortium Chicago (USA), 1996, pp.287-300.

6. B. Steffen, T. Margaria, A. Claßen: *Heterogeneous Analysis and Verification for Distributed Systems*, "Software: Concepts and Tools" Vol.17(1), pp.13-25, March 1996, Springer Verlag

7. B. Steffen, T. Margaria, A. Claßen, V. Braun, M. Reitenspieß: *"A Constraint-Oriented Service Creation Environment,"* Proc. PACT'96, Int. Conf on Practical Applications of Constraint Technology, 24-26 Apr.1996, London (UK).

8. B. Steffen, T. Margaria: *Method Engineering for Real-Life Concurrent Systems*, position statement, ACM Worksh. on *Strategic Directions in Computing Research*, Working Group on Concurrency, ACM Computing Surveys 28A(4), Dec. 1996, http://www.acm.org/surveys/1996/SteffenMethod/

9. B. Steffen, T. Margaria: *Tools Get Formal Methods into Practice*, position statement, ACM Worksh. on *Strategic Directions in Computing Research*, Working Group on Formal Methods, ACM Computing Surveys 28A(4), Dec. 1996, http://www.acm.org/surveys/1996/SteffenTools/

10. B. Steffen, T. Margaria, A. Claßen, V. Braun: *Incremental Formalization: a Key to Industrial Success*, in "Software: Concepts and Tools", Vol.17(2), pp. 78-91, Springer Verlag, July 1996.

11. B. Steffen, T. Margaria, A. Claßen, V. Braun, M. Reitenspieß, H. Wendler: *Service Creation: Formal Verification and Abstract Views*, Proc. 4th Int. Conf. on Intelligent Networks (ICIN'96), Nov. 1996, Bordeaux (France), pp. 96-101.

12. B. Steffen, T. Margaria, V. Braun, N. Kalt: *Hierarchical Service Definition*, Annual Review of Communic., Int. Engineering Consortium, Chicago, 1997, pp.847-856.

13. B. Steffen, T. Margaria, A. Claßen, V. Braun: *The METAFrame '95 Environment*, Proc. CAV'96, Juli-Aug. 1996, New Brunswick, NJ, USA, LNCS 1102, pp.450-453, Springer Verlag.

The Generation of Service Database Schema through Service Creation Environment

Jeomja Kang, Jeonghun Choi, Sehyeong Cho

Electronics and Telecommunications Research Institute (ETRI)
161 Kajong-Dong, Yusong-Gu, Taejon, 305-350, KOREA
Tel : +82-42-860-4880
FAX : +82-42-861-2932
E-mail : (jjkang, jhchoi, shcho)@dooly.etri.re.kr

Abstract. For rapid and easy service development and deployment, the concept of Service Creation Environment (SCE) has emerged. The SCE helps creation of the service logic and the service database schema by using automated tools. The service logic and service database schema creation is based on SIBs of ITU-T CS-1. In this paper, we show the design and implementation of our SCE system. And also, we describe the service database design process for database creation and the database information translation process, which is one of the most difficult part in service creation process.

1 Introduction

The concept of Service Creation Environment (SCE) has emerged as a solution for rapid development of network service software. Traditionally, network service software has been implemented according to the conventional software development process - requirements, design, code, and test - and this process requires several years. But, the time of this process may now be dramatically reduced with the introduction of Service Creation Environment [1]. With an SCE, a service designer designs service by combining standardized and modularized components that represent network capabilities in abstract. Network service software is meant to be generated by using tools that translate the service logic into actual service logic programs that are practically executable on top of the given service control platform.

When service logic is deployed into the network, service database must be created through Service Management System (SMS). For this purpose, we considered the environment for service database design in SCE and then we reflected this into the prototype of SCE. In service data description area, Data Definition Language(DDL) is widely used in many SCEs[2,3]. DDL is an SQL like language for defining typed data and the schema of database used by the SCP to hold service data. However, it is difficult to understand the DDL and to extract the service data from service requirements.

In this paper, we show the design and implementation of our SCE system and tools, which constitute SCE. We also describe the service database design process by using automated tools to help generate service database schema applicable to RDBMS.

In section 2, design and implementation of our SCE is described. In section 3, We describe the service database schema generation method and database information translation method. In section 4, we conclude this paper.

2 Service Creation Environment

2.1 Characteristics of SCE

In our SCE, we describe the service by using visual programming environment at first and then we create and test the service logic. Our SCE is based on UNIX system, so our SCE is independent of specific hardware. The characteristics of our SCE are as follows.

- *Integrated service creation environment* : Our SCE is composed of several tools, which are tightly integrated for service creation according to consistent service creation method, which is independent of the service.
- *Support specification of ITU-T CS-1*: Our SCE supports ITU-T CS-1 and application program interface (API) of Korea Telecom, which is our target platform. The description of service by using graphic environment is called the GSL (Global Service Logic) and GSL is checked the correctness of syntax and semantic. So, users can detect errors easily.
- *Support service database design* : Our SCE support service database design. We can extract service data from SIB parameters of GSL and then enhanced or modified as necessary by using a graphical database design tool.
- *Validate service logic* : For distributed service logic validation, we use a SDL environment and simulate the distributed service logic created automatically from GSL.

2.2 Service Creation Process

Our service creation process and methodology are based on Intelligent Network Conceptual Model (INCM). The INCM is structured into four planes: the Service Plane (SP), the Global Functional Plane (GFP), the Distributed Functional Plane (DFP), and Physical Plane (PP). At the service plane, we see services and service features. A service is composed of service features. In this plane, we define services and service features from a user's viewpoint. At the global functional plane, we model the service from global viewpoint. The services are realized by combining SIBs of ITU-T CS-1. At the distributed functional plane, we validate the service logic whether it is feasible or not. At the physical plane, we create the service logic program, which is executable in the target platform.

2.3 Configuration of SCE

Based on our service creation process, our SCE system is composed of four functions as in Fig. 1. The service logic design function defines the service logic and checks the syntax of SIB by using graphic user interface. The service database design function generates the service database schema, which is needed for execution of service logic by user. The service validation function checks and validates the described service logic whether it is feasible or not through simulation method. The code generation function creates the service logic program, which is executed in the target platform. These functions are integrated through common representation.

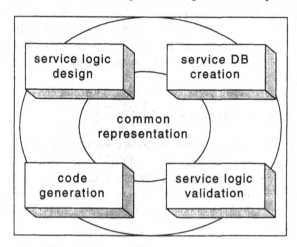

Fig. 1. Configuration of functions in our SCE

The tools for implementation of SCE functions are composed of global service logic editor, common interchange format processor, distributed service logic translator, distributed service logic simulator, service database creator and code generator.

The Fig. 2 shows configuration of our SCE tools and platform.

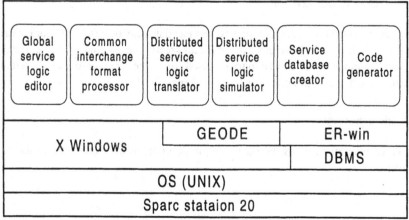

Fig. 2. Configuration of tools and platform in our SCE

(1) Global service logic editor

The global service logic editor supports service description by combining standard ITU-T CS-1 SIBs. The service designer uses this tool to define the global service logic of IN. This tool provides the graphical user interface for syntax checking. The service logic described by this tool is saved to CIF(Common Interchange Format) file. We use standard OSF/Motif graphical user interface.

(2) Common interchange format processor

The common interchange format processor stores and extracts common data and used by other tools of our SCE. This tool does not have direct interaction with the user of SCE system. Common Interchange Format (CIF) is defined by Backus Normal Form (BNF). This tool is also responsible for constructing the Abstract Syntax Tree (AST) for a given CIF file.

(3) Distributed service logic translator

The distributed service logic translator creates the Distributed Service Logic (DSL). This tool is used for validation of given global service logic. DSL translator maps SIBs of GSL into SDL diagrams that embodies Functional Entity Actions(FEAs) that are related to information flows between functional entities in ITU-T recommendation Q.1214. The mapping between SIBs and FEAs is defined in ITU-T recommendation Q.1214.

(4) Distributed service logic simulator

The distributed service logic simulator validates the service logic completed by global service logic editor. The service designer guides the simulation of the logic through a graphical interface.

(5) Service database creator

The service database creator supports service database design. This tool extracts service data from SIB parameters of GSL and creates the schema for target database system. A commercially available tool is used for further engineering of the database schema. (We actually integrated ER-win by Logic Works.)

(6) Code generator

The code generator creates the service logic program which executable in the target platform.

3 Service database design process

3.1 Mapping of the SIB parameter

Log call information SIB, screen SIB, service data management SIB, status notification SIB and translate SIB are related to database access in Q.1213 SIB set of ITU-T. Each SIB performs read and writes function for specific data in specific file or database table. We can find two types of information from the SIB parameters. First, we can extract service data for database structure. The SIB parameter structure is mapped into table structure of database in relational database, that is, the file indicator of service support data in SIB parameters is mapped into entity of table in database and data requesting read or write is mapped into attribute of table. Table 1 shows the mapping of SIB parameters of SDM into database table. In SDM SIB, if action is data retrieval then we retrieve data, which specify element indicator of file indicator. That is, file indicator of SIB parameter is mapped into table name in relational database and an element indicator of SIB parameters into attribute in relational database.

Second, we can provide distributed service logic and code generator with the detailed information related to database access for service logic program creation. For providing detailed information, we map SIB parameters of Q.1213 into API parameters of specific platform. As the result of the mapping, we can extract only file indicator and field name in a specific platform. So, we must redefine detailed information such as data type, data length, key value that actually requires relational database. Table 1 shows required information for service logic program creation in a specific platform. In the third column of Table 1, SIB-File-Indicator is mapped from file indicator of Q.1213. Since SIB-File-Indicator represents the GSL user's viewpoint, we use the table name, which actually requires entity of database table. The modification field is used for marking service data update state. If service data are updated then we can extract only the service data that are marked as modified.

Table 1. SDM SIB parameter mapping

SDM SIB of Q.1213	database table	API of specific platform
Service Support Data		
- File indicator	Entity	*SIB-File-Indicator*
- Action		*Modification*
- Element indicator	Attribute	Table-name
- Inc/Dec value		Key-field-name
- CIDFP-Info		Key-field-type
- CIDFP-Element		User-key-value
- CIDFP-Retrieve		SIB-Element-Indicator
- CIDFP-Error		*Field-name*
		Field-type
		Field-size

3.2 Generation of the service database schema

The service database schema is used for service logic program creation and for service specific database table creation. Currently the DB schema generation is targeted for our target platform, which uses the Oracle RDBMS, but the tool can support a dozen of different Database systems.

For service database schema implementation, we use commercial database design tool supporting relational data model. The commercial database design tool supports creation of various relational database schema.

The service database schema contains the definition of tables and the relationship among tables. Fig. 3 shows the generation process of the service database schema.

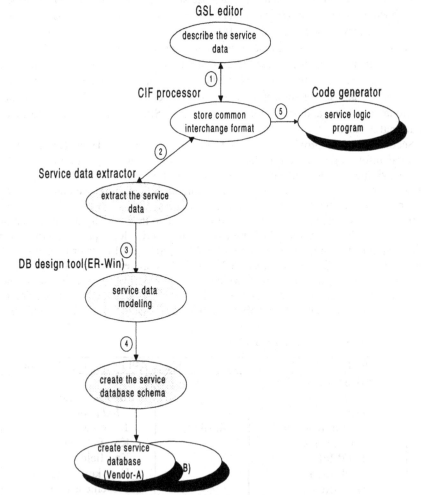

Fig. 3. Generation process of the service database schema

(1) Describe the service data

The service data described by using the global service logic editor is based on SIB parameters of ITU-T Q.1213. In service data description, the users are classified into two classes. One class is users who does not have detailed knowledge about database organization and describes only SIB parameter. The other class is users who have detailed knowledge about database organization and describes detailed information of specific platform. This is stored Common Interchange Format (CIF).

(2) Extract the service data

This step is for selection the service data for a specific service. Because the GSL is described by using the global service logic editor has enough information about the service data, we can extract the service data by using automated tool. The service data extraction stage is composed of four steps. First, the service data extractor requests Abstract Syntax Tree (AST) to CIF processor. Second, the service data extractor selects SIBs from AST related to the database access, i.e., screen SIB, service data management SIB and so on, in GSL. Third, service data extractor extracts elementary service data, which are needed to fulfill the service from selected SIBs. These data can be easily translated into attributes. Fourth, if service data are extracted, then MODIFICATION bit reset to FALSE. When service data are described, MODIFICATION bit is set to TRUE and is used for marking service data update state. The following excerpt shows how the service data are extracted. If SIB_FILE_INDICATOR and FIELD_NAME parameter of service data management SIB in GSL are described then we can extract service data from SIB_FILE_INDICATOR and FIELD_NAME of SDM SIB parameter. SIB_FILE_INDICATOR maps into table name and FIELD_NAME maps into column name of RDBMS.

```
[Service Data Management SIB ]
SIB_SERVICE_DATA_MANAGEMENT  1(SDM SIB number)
    SIB_FILE_INDICATOR          = "abstract file indication"
          MODIFICATION          = TRUE
          TABLE_NAME            = "pcs_profile"
          KEY_FIELD_NAME       = pcs_number
          KEY_FIELD_TYPE       = SIB_FIELD_STRING
          USER_KEY_VAL         = "called_party_number"
    SIB_ELEMENT_INDICATOR
          FIELD_NAME           = screeen_passwd
          FIELD_TYPE           = SIB_FIELD_STRING
          FIELD_SIZE           = 4
    ACTION                = SIB_SDM_RETRIEVE
    CIDFP  VARIABLE_NAME       = pcs_number
    IF SIB_SUCCESS THEN    2 (next SIB number)
    END;
```

If a user describes the service data indicated in bold font, we can extract service data from SDM SIB example and the extracted is as follows. The format of the output file is ".sql" format. The *abstract file indication* extracted from *SIB_FILE_INDICATOR* is renamed into an appropriate table name in the service data-modeling step. CREATE TABLE is generated when TABLE_NAME has different name. After checking the attribute name of its uniqueness, the attribute with same TABLE_NAME is inserted only when attribute has different name.

CREATE TABLE abstract file indication
(
       screen_passwd   CHAR(30) NOT NULL
);

If all parameters of SDM SIB in GSL are described, we can extract the following service data from SDM SIB example. If *SIB_FILE_INDICATOR* and TABLE_NAME are defined then TABLE_NAME has priority.

CREATE TABLE pcs_profile
(
       pcs_number   CHAR(30) NOT NULL,
       screen_passwd CHAR(4) NOT NULL
);
ALTER TABLE pcs_profile
ADD(PRIMARY KEY(pcs_number));

In this example, the extracted service data from SIB parameter of GSL are influenced by service data description method: If the user describes the service data in detail, then we can extract more information.

(3) Service data modeling

This step is to define the characteristics such as data type, data length, key, and attributes, in case new attributes are needed, when they are added in an interactive way of service data extracted in the previous step and to perform data modeling by using commercial tool. The extracted service data by using reverse engineering function of DB design tool are used as the input file for the tool.

(4) Create the service database schema

This step is to generate the service database schema dependent on the target platform by using the relational tables of service data modeling step. Our SCE creates various database schema of RDBMS(Relational DataBase Management System) such as DB2, Ingress, SYBASE and so on. Our SCE creates schema for Oracle RDBMS by default, since the current target platform uses Oracle RDBMS. The service database schema is created either in table format of target database or in a sql script file. If the service data update of GSL occurs, then we create a sql script file and reflect updated service data

(5) Create the service logic program

After service database schema is created completely, we can describe detailed information by using global service logic editor. Detailed information means database information related to database access, i.e., table_name, data_type, data_length and so on. CIF processor stores described detailed information and then code generator uses this detailed information for service logic program creation. We describe method of detailed information translation in the section 3.3

Fig. 4 shows the service database design process using ER-win interface. At first, the service data is extracted from SIB parameters of GSL by using automated tool. The extracted service data by using DB design tool are used as the input file for the tool. The attribute name is redefined and relationship between tables is defined.

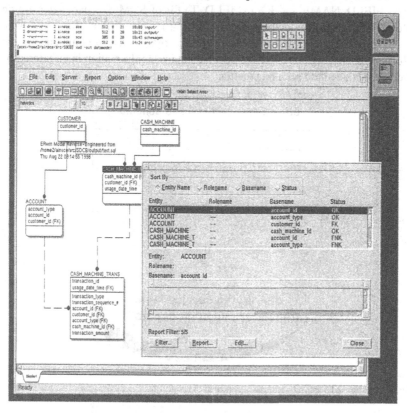

Fig. 4. ER-win interface for service database schema

3.3 Translation of database information

In order to help mapping abstract parameters in GSL to detailed information in DSL (Distributed Service Logic), database information provide distributed service logic translator and code generator with the detailed information related to database access. The SIB parameters and SIB values in GSL are abstract information. Therefore it is difficult to translate the database access information automatically. CIF processor stores detailed information filled in by using global service logic editor. In our SCE, we allow the user to supplement missing information in an interactive way. In the future, we will consider database information translation by automated tools. We classify the user according to database knowledge. If a user does not have detailed knowledge about database organization, we allow this user to describe only SIB parameter defined in ITU-T recommendation Q.1213. If a user has detailed knowledge about database organization, we allow this user to describe detailed information related to database access. In the latter case, database information translation is not required. Fig. 5 shows the database information translation process. For example, if service data management SIB in GSL is described, then we can provide detailed information by using global service logic editor. TABLE_NAME, KEY_FIELD_NAME and KEY_FIELD_TYPE of SDM SIB parameter are detailed information related to database access. But, it is very difficult for a user who does not have detailed knowledge about database organization to complete the database information. So, we complete the detailed database information in an interactive way after service database schema creation.

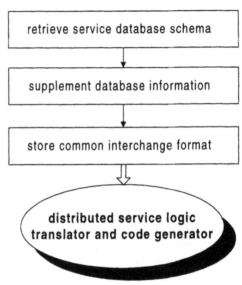

Fig. 5. Database information translation process.

3.4 Reflection of Service data update

Normal cases of service database schema creation for database organization and database information is described in 3.2 and 3.3 section. But if service logic and service data update is occurred, then we must reflect it into created service database schema and service logic program generated already. Figure 6 shows reflection of service data update.

We assume normally described service logic GSL′ and service database schema DB′. If GSL′ is updated, we can create GSL″ adding updated information α. A user opens the GSL′ in CIF and updates service logic and service data of GSL′. When service data of GSL′ is updated, modification bit of SIB parameter related to database access set to TRUE (1). Newly described service logic GSL″ is saved as CIF (2) and service data extractor extracts service data only when modification bit is TRUE (3). After service data extraction, modification bit is set to FALSE (4). Finally, newly extracted service data are used as the input in database design tool (5) and already created DB′ is merged with service data extracted newly by the user (6). The service database schema DB″ that is executed in target database system (7) is created from the merged service data file. A user updates the detailed information by using global service logic editor and saves CIF (8). The code generator creates the service logic program (9) from GSL″ adding to detailed information.

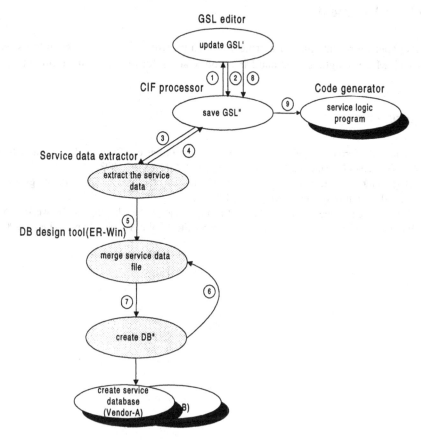

Fig.6. Reflection of service data update

4 Conclusion

In this paper, we showed the design and implementation of our SCE system. We also described tools consisting of our SCE. We focused on the service database design process. Service database design process through our SCE is composed of service database schema generation and database information translation. Service database schema generation creates the service database schema. Database information translation provides the distributed service logic translator and the code generator with the detailed information related to database access.

Service database table creation is made easier by using the service database schema generator and the database information translator to handle data related aspects of service creation, which is one of the most difficult parts in service creation process.

Acknowledgment

This paper represents part of a project funded by Korea Telecom. The authors wish to thank Ms. Youngmee Shin and Mr. Choongjae Im for their helpful comments.

References

1. Richard A. Orriss, Service Creation-Making the Promise a Reality, IEEE International Conference onCommunication'93, pp.1531-1533, May 23-26, 1993.
2. Marj syrett, David Skov, and Anders Kristensen, HP in IN, in proceedings of IN Workshop, Ottawa-Canada, May 9-11, 1995.
3. Gilles Bregant and Roberto Kung, Service Creation for Intelligent Network, XIII International Switching Symposium, pp. 45-50, Stockholm-Sweden, May 27-June, 1990.

A Study of Intelligent Multimedia Services over PSTNs and the Internet

L. Orozco-Barbosa[1], D. Makrakis[2] and N.D. Georganas[1]

[1]Multimedia Communications Research Lab
University of Ottawa
161 Louis Pasteur
Ottawa, Ont. K1N 6N5 Canada

[2]Advanced Communications Engineering Centre
University of Western Ontario
Engineering Science Building
London, Ont. N6A 5B9 Canada

Abstract. The Intelligent Network (IN) architecture promises easy development and fast deployment of sophisticated communication services. In this paper, we present the results of our research in the area of Intelligent Multimedia Services. One of the main results has been the evaluation of novel multimedia services operating over different network platforms. One of the new services being designed is a Multimedia News-on-Demand application allowing users to browse through items of current affairs and select a specific article for viewing. This service is supplemented by an Intelligent Multimedia Gateway providing interconnectivity between the public service telephone network and the Internet. The study of this service has proved useful in gaining insight on the way novel intelligent multimedia services across network platforms will benefit from the use of IN technology. A performance evaluation of the service has also proved valuable to identify the major system and traffic parameters to consider when designing and deploying IN multimedia services across multiple networks.

1 Introduction

The Intelligent Network (IN) architecture[1][2][3][4], and now the Advanced Intelligent Network (AIN) architecture[5], is an emerging technology that promises easy development and fast deployment of sophisticated communication services. Such an advancement over the traditional telephony system is largely attributed to the concentration of "intelligence" on new nodes called Service Control Points (SCPs). Users are linked to the network via local exchanges (LE's). Ordinary phone calls that do not involve IN services are handled solely by Service Switching Points (SSPs), which are mostly traditional switches. Whenever an incoming dialed number is detected by the *triggers* at the SSP as an IN service number, a query message is sent from the SSP to the SCP via the Common Channel Signaling Network, in particular, using the Signaling System 7 protocol (CCS7). The SCP exchanges CCS7 messages with the SSPs that require instructions for the establishment of the connection. Eventually, the SSPs are responsible for establishing the voice channel for the call. See Figure 1 for reference. If multiple SCPs are reachable from each SSP in a CCS7 network, one or more layers of intermediate switches called Signaling Transfer Points (STPs) can be inserted between SCPs and SSPs.

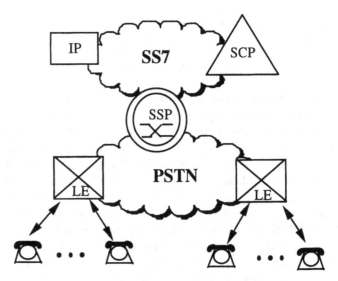

Fig. 1. Components of the Intelligent Network

The financial benefits obtained by the use of IN are considerable[6]. For this reason, most of today's major R&D organizations and providers of communication services, such as Bellcore[2], GTE[7], AT&T[8], Alcatel[9], Siemens[10], NTT[11], Nortel[12], Ericsson[13], are involved heavily and moving aggressively in the IN field. At the same time, the major players in the communication industry, are actively participating in the definition of international standards[3][4].

In terms of future services, two seem to be the main factors that should be taken seriously into consideration in the design of the future IN technology. The first is the introduction of multimedia (MM) communications services[14]. Multimedia applications will have a considerable impact on our professional, social and personal functions, and they will affect considerably a number of our industries, such as medicine (telemedicine), education (distance education) or marketing (telemarketing). Future IN technology should use the various multimedia applications as tools, and combine them to generate new customer service applications. Examples of such potential services is the creation of an Integrated National Medical Distributed Multimedia Database (INMDMD), which will allow universal access to medical records and medical research activities all across the country based on the use of personal identification numbers.

The second area which will require the service of the IN technology is the interconnectivity of a wide variety of communications networks, ranging from the Internet to Personal and Mobile Communications (PMC) networks. The communications industry has already invested considerable resources in R&D for a multitude of networks over the past several years. Today, there are expectations and projections of a booming market: only the PMC industry itself continues to grow with the rapid rate of 25% annually. Early in 1994, the personal communications industry association in the United States forecasted a triple-digit growth in the number of subscribers to PMC services during the ten coming years. By the year 2003, it is expected that, in the U. S. A. only, 164 million subscribers will have joined services such as personal communications, cellular radio, mobile satellite services, paging, etc.[15].

In this paper, we examine the use of IN technology for the design and performance analysis of multimedia services across network platforms. We start by analyzing the principles to implement intelligent services across different platforms. We illustrate these principles by undertaking the design and performance analysis of an Intelligent Multimedia News-on-Demand supported by a Multimedia Gateway service for interconnecting the PSTN to the Internet. This service allows the retrieval of multimedia documents from databases in the Internet through the PSTN. The development of this service is based on the Multimedia News-on-Demand[16] application and Multimedia Gateway[17] service developed at the Multimedia Communications Research Lab (MCRLab) of the University of Ottawa. We also look at the performance issues affecting the performance of this kind of service. Some numerical results are also given.

2 An Intelligent Multimedia News-On-Demand Service

2.1 System Description

Figure 2 depicts the general architecture of a system enabling PSTN users to retrieve of multimedia news from multimedia databases in the Internet. It is based on the Multimedia Gateway and Multimedia News-on-Demand applications developed in our lab at the University of Ottawa[16][17]. This gateway application was originally developed to provide the means of offering fax services to Internet users and Internet services to multimedia fax users whose workstations are equipped with the new BFT (Binary File Transfer) fax interfaces [17]. In the case of the Multimedia News-on-Demand application, the main role of the Multimedia Gateway will be to provide the proper interface to support the Multimedia News-on-Demand application across network platforms.

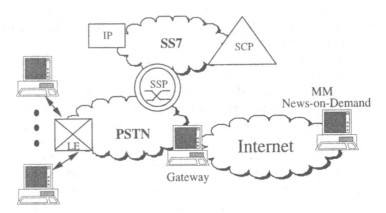

Fig. 2. System Configuration

The Multimedia-News-on-Demand application is a specific presentational application. The application allows users to browse through items of current affairs and select a specific article for viewing. The users can interact with the presentation of the document in the same way as they would when viewing tapes from a VCR. The articles are stored in a remote multimedia database, and they are composed of medias such as text, audio and video. Through multiple servers, activities such like multimedia data retrieval and document delivery are performed continuously in real time.

The IN technology can enhance the multimedia News-on-Demand service by dynamically addressing communication entities based on the callers' attributes, callers' equipment, time-of-day, services, and so on. The proposed service takes the advantages of the IN technology, and provides a value added service to PSTN users. Potential users of this type service are industries such as tourist agencies, hotels, resorts, restaurants, etc. The operation of the proposed service is as follows:

- the user picks up and dials the "universal" IN multimedia number. This can be a Freephone number or a Premium number (the cost of the service is charged to the calling party). As soon as the SSP recognizes that this call requires special handling, it passes the information to the SCP along with the calling number.
- the SCP consults the appropriate database to determine the area from where the call is originated. The SCP connects the calling party to the appropriate IP. At this point, there are two possibilities. The IP might be using speech recognition, or the dial-key pad as means of providing input to the system. The system asks the caller, among other things, the type of equipment she uses. This will allow to determine the protocols being used all along the duration of the call.
- once having identified the service requested, the SCP directs the user to a more specialized IP in charge of getting the information such as the topics of interest, e.g., sports. The main responsibility of this IP will be to collect all the information needed to look for and retrieve the information required by the user.
- the SCP is required to look for the gateway to be used for the call. This task is

needed for the processing of the type of service being proposed and it can be defined in a generic way for any IN service requiring to establish a connection to the Internet. Two of the main issues in choosing a gateway are the location of the gateway and the load distribution among the various possible gateways. In the former case, the closest gateway to the user's location will be the natural choice to avoid any long distance charges. These two factors, the location and load distribution, will determine the cost as well as the quality of the service to be provided to the user.

- once a gateway is chosen, the SCP instructs the SSP to establish a connection between the IP and the gateway. The IP takes in charge the task of connecting to the multimedia database in the Internet and request the desired information. The user is informed that her request has been placed.

- once that the required multimedia News-on-Demand has been contacted, a connection is established between the gateway and the user. The proper dimensioning of the resources at the gateway the multimedia will provide the baseline for assuring the quality of service required by the service. This is the main aim of the study presented in Section IV.

The main objective of this kind of facility is to provide to the PSTN users access to the Internet or any similar network wide range of services. IN is the technology enabling the creation of the universal (unique) interface to the global network.

2.2 Application Requirements

In this paper, our main objective is to study the major performance issues to consider for the deployment of efficient IN multimedia services. One of the major characteristics of distributed multimedia applications is the spatio-temporal integration of multimedia objects[14]. Multimedia integration can be described by scenarios, that represent the temporal (time domain) relationships among the data objects (space domain) involved in the multimedia objects. In a previous paper[16], we have analyzed the requirements to support continuous media over packet-switched networks. One of our main results has shown that the number of intermedia synchronization errors is enormously reduced by increasing the buffer capacity at the receiving end. In the case of an interworking system interconnecting a packet-switched network to a circuit switching network, the gateway will be in charge of compensating of any synchronization errors. The gateway's ability to properly carry this task will depend on the number of trunks used to interconnect the two networks and the buffer size to compensate the loss of synchronization under various load conditions. In order to be able to evaluate the effectiveness of the system, we must define the proper performance metrics of interest. The main quality of service parameter of interest is the call blocking probability. For traditional phone calls, call blocking is defined as the ratio between the number of call connection requests refused due to lack of resources, telephone lines, and the total number of call connection requests. In the case of IN services, the procedures involved in the IN service provisioning operations demand the deployment of facilities not found in traditional PSTN. IN calls require the use of specialized units able to handle the service processing requirements. In this case, IN may be blocked due to lack of insufficient resources, such as memory or enough processing power, needed to handle the service provisioning operation in a timely basis.

We undertake the performance evaluation of the proposed IN multimedia News-on-Demand service by stressing the major requirements of this service as compared to other IN services. We have chosen to focus in analyzing the effects on the performance of the IN service due to two of the major features of this kind of IN service, namely: stringent processing requirements and the interconnection of two network platforms. In the following section, we present the simulation model of the system configuration under consideration.

3 Performance Evaluation

3.1 Simulation Models

Figure 3 depicts the simulation model of the system configuration under study. The model consists of three queuing centers, three delay centers and the model of the gateway. The SCP, IP and SP are represented by queuing centers. Each queuing center consists of single server and associated queues. The SS7 network, the Internet and the database server are modeled by delay centers. For clarity reasons, the SS7 center is illustrated four times in the figure. The details of the queuing model of the gateway are shown in Figure 4.

In the actual system, when a new call process is initiated, it goes first to the LE, and from there it is forwarded to the SSP. There is some time elapsing between the moment the call is initiated and the moment where the SSP receives the call request. This is the combined delay due to the propagation time through the lines, as well as the waiting and processing times at the LE.

The traffic entering the various queues of the SSP is generated by a variety of services. As we are interested in evaluating the performance of the IN service, we only examine the behavior of the MM-News-on-Demand service as it progresses through the network. We do this by allowing only the multimedia IN calls , referred from now on as MM-calls, to go through all the network elements involved during the execution of the service. The numbers of interactions between the IP and the user is an important parameter of the service. Each interaction does not only take processing time from the IP, but it also generates additional traffic over the signaling network.

The main functions of the SSP considered in our model are: the processing of new call requests and the processing of calls having previously being forwarded to either the IP or the SCP. Three different types of calls will be considered: emergency, regular phone calls, IN calls and multimedia IN calls. The SSP model consists of four different queues. The first queue, referred as emergency calls, collects emergency calls which can never be rejected. The second queue, SCP responses, handles the ongoing calls which are waiting for further call processing based on the SCP instructions. A background load of responses from the SCPs has been introduced to take into account the responses generated from other on-going IN calls as well as the SCP responses originated for the processing of the IN multimedia call. The third queue, IP responses, contains the exchange of information between the users and the IPs. Similarly to the previous queue, background as well as response related to the reference IN call are considered. The fourth queue, new calls, contains the requests of new calls, both IN and non-IN are modeled by this queue. The parameters of traffic are defined as follows: λ_{SSP-E}, $\lambda_{SSP-SCP}$, λ_{SSP-IP} and $\lambda_{SSP-NEW}$ are the arrival rates for emergency calls, calls waiting SCP instructions, calls waiting user interactions and new calls, respectively.

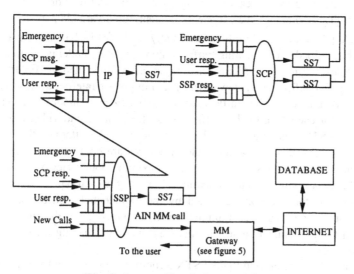

Fig. 3. System Simulation Model

The SCP model is built in a similar way with the SSP model (see Figure 3). The main functions of the SCP being modeled are: accepting SSP requests and processing them, activating IP to perform the function of user interactions, receiving user information collected by the IP and sending the call processing information to the SSP. The model consists of three queues to handle emergency calls, SSP requests and replies from the IP to be directed to the users. The parameters of traffic load to each queue are defined as follows: λ_{SCP-E}, $\lambda_{SCP-SSP}$, λ_{SCP-IP} are the mean arrival rates of emergency calls which need SCP processing, SSP requests and incoming user responses from IP, respectively.

The IP is a system which controls and manages functions such as voice synthesis, announcements, speech recognition, and digit collection. Its main simulated functions are: accepting messages from the SCP which inform the IP about the resources or procedures to be used on the call, connecting the caller through the SSP, prompting users and collecting user responses and returning the user input to the SCP. We assume that emergency calls are directly connected to human operators rather than to intelligent peripherals. Consequently, there is no emergency call processing in the IP node. Two external traffic generators are used to model the overall traffic load of two different types in the IP node. The parameters of this queuing center are defined as follows: λ_{IP-SSP}, λ_{IP-SCP} are the mean arrival rates of user responses (through the SSP) and SCP messages, respectively.

The SS7 network is the underlying transport mechanism between the SCP and SSP, and between the SCP and IP. From the application point of view, the SS7 network receives a message from a network node (the SSP, SCP or IP), and delivers this message to another network node after some delay. The amount of the delay depends on the traffic load of the SS7 network. In our simulation, we model the delay, t_{SS7-d}, that is experienced by a packet traveling through the SS7 network as the sum of one fixed part (T_c) and a negative exponentially distributed variable part (t_v). This model is accurate when the signaling links are lightly loaded[18]. To study the performance under a heavy loaded of signaling traffic, a more complex model is required.

The gateway model used in our simulation is illustrated in Figure 4. Upon connecting to the SSP, the gateway will perform proper Internet access, based on the messages of the SSP. The traffic load of the gateway is modeled by using traffic generators for each type of the traffic, signal, text, image and video traffic, on both networks: PSTN and Internet. Each traffic arrival pattern follows a Poisson distribution. The text, image and video file sizes are conformed to the exponential distribution, but we impose minimum and maximum boundaries on the file size, that is, we choose a minimum and a maximum file size for each type of files: text, image and video. If the file size generated by the exponential distribution generator is smaller (or greater) than the minimum (or maximum) value, the file size takes the minimum (or maximum) value. In this way, we avoid having a zero or extremely large file sizes.

The gateway is connected to the PSTN through groups of 24 telephone lines, and to the Internet through several T1 lines. However, only a group of lines and a T1 have been explicitly modeled (see Figure 4). We have considered the use of a group line of 24 lines each having a nominal service rate of 64K bits per second corresponding to a T1 level in the hierarchical structure of the telephone network in North-America. The T1 line is modeled as one server with the service rate of 1.544 Mbits per second. We take into account the effect of the traffic generated by other groups of lines and T1 lines by including several traffic load generators of various types going into the gateway. The traffic related to these background generators are fed into the group of lines and T1 line explicitly modeled. It is assumed that upon receiving the file associated to one of these generators, the gateway processes it and then directs it to an output telephone or T1 lines other than those explicitly modeled. This procedure allows us to evaluate the performance of the system under different traffic loads.

The gateway is connected to the PSTN through groups of 24 telephone lines, and to the Internet through several T1 lines. However, only a group of lines and a T1 have been explicitly modeled (see Figure 4). We have considered the use of a group line of 24 lines each having a nominal service rate of 64K bits per second corresponding to a

T1 level in the hierarchical structure of the telephone network in North-America. The T1 line is modeled as one server with the service rate of 1.544 Mbits per second. We take into account the effect of the traffic generated by other groups of lines and T1 lines by including several traffic load generators of various types going into the gateway. The traffic related to these background generators are fed into the group of lines and T1 line explicitly modeled. It is assumed that upon receiving the file associated to one of these generators, the gateway processes it and then directs it to an output telephone or T1 lines other than those explicitly modeled. This procedure allows us to evaluate the performance of the system under different traffic loads.

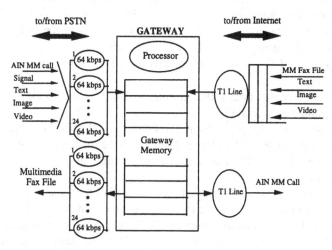

Fig. 4. Model of the Multimedia Gateway

Upon receiving a request for the retrieval from the PSTN, named AIN MM-call in Figure 4, the gateway issues a request through the T1 interconnecting the gateway to the Internet. The processing time of this request at the gateway is assumed to be negligible. Once the request has been processed by the database server, the multimedia file is sent to the gateway. The gateway operates following a store-and forward scheme, i.e., the gateway receives the various files or service requests from both the PSTN and the INTERNET, stores them in the memory for protocol translation before sending them into the INTERNET or PSTN. Call blocking may occur for two reasons: i) all transmission servers are busy when the IN MM call arrives; or ii) lack of memory in the gateway for storing the multimedia file the IN MM call.

We model the Internet and the database server as two delay nodes. The delay experienced by a message while traveling through the Internet or processed by a database server residing in the Internet is modeled by a random number which statistically follows a fixed value, T_{INT} for the Internet and T_{DB} for the database server, and an exponential distribution with a mean value, t_{INT} for the Internet, and t_{DB} for the database server.

3.2 Simulation Results

Based on the models presented in the previous section, this section presents the simulation results of the IN multimedia Video-on-demand service. The simulation model has been built using OPNET, a powerful simulation tool for tele-communications[19]. The main aims of this study are twofold. First, we are interested in evaluating the impact on the performance when implementing IN services requiring intensive processing functions, i.e., the use of SCP, IP and related logic. This is one the most important factor in the deployment of IN multimedia services. Second, the study of the impact of various kinds of multimedia traffic mixes over the gateway providing the interconnectivity between the PSTN and the Internet. The efficient operation of the gateway is central to the overall success of the system in providing the quality of service required by IN multimedia services.

Bearing in mind the two aforementioned objectives, we have made the following important assumptions:
- all the traffic arrival rates follow a Poisson distribution, and all the service times are distributed exponentially.
- 50% of the total calls presented to the PSTN are IN calls, i.e., calls requiring the services of the SCP, 50% of these IN calls involve the use of the IP services. From the overall arriving IN calls a certain percentage, given by P_{MM} x 100, are multimedia calls.
- the traffic arrival rates and service times of the SSP, SCP and IP have initially been set to fix the utilization of these three servers to around 80 % of the nominal server capacity. This is common practice when evaluating switching and processing elements in telecommunications and allows to study the impact when implementing services requiring more and more the services of the SCP and the gateway.
- the use of digital video has been considered. Digital video is becoming an important vehicle of communication[14]· The study of the requirements to support video communications is essential for the deployment of effective IN multimedia services.

The basic set of parameter values used for the SSP, SCP and IP is listed in Table 1. Both the arrival rates and service times of the three servers are given. The service times of the servers, denoted by x_{scp}, x_{ssp} and x_{ip}, are assumed to be the same for all classes of transactions. This set of parameter values will be referred from now on as the *basic value set*. The number of user interactions with the IP is one of the important parameters to consider and it has been varied in each simulation. This number refers to the number of requests/replies between the IP and the user each time a trigger is activated. For each simulation, this number follows a normal distribution with a fixed mean value for the entire simulation time. We have used the mean number values ranging from 2 to 6.

Table 1. Arrival rates and service times for SSP, SCP and IP

SSP	SCP	IP
$\lambda_{SSP-NEW} = 100$ calls/sec	$\lambda_{SCP-E} = 0.1$ calls/sec	$\lambda_{IP-SSP} = 50$ calls/sec
$\lambda_{SSP-SCP} = 50$ calls/sec	$\lambda_{SSP} = 50$ calls/sec	$\lambda_{IP-SCP} = 25$ calls/sec
$\lambda_{SSP-IP} = 50$ calls/sec	$\lambda_{SCP-IP} = 25$ calls/sec	
$\lambda_{SSP-E} = 0.1$ calls/sec		
$x_{SSP} = 1/\mu_{SSP} = 3.5$ ms	$x_{SCP} = 1/\mu_{SCP} = 10$ ms	$x_{IP} = 1/\mu_{IP} = 10$ ms

The gateway service rate is 25 Mbits/sec and the gateway memory is 2 Gbytes. The multimedia file size retrieved from the database server follows an exponential distribution with mean 400 Kbytes. The statistics of the background traffic arriving to the gateway are given in Table 2.

The Internet node service times have been set to $T_{INT} = 50$ ms and $t_{INT} = 50$ ms. The service times of the Database server in the Internet are $T_{DB} = 50$ ms and $t_{DB} = 50$ ms and those of the SS7 network are $T_C = 10$ ms and $t_V = 1$ ms. These values have been taken from reference[18].

Table 2. Multimedia Traffic Statistics

Traffic type	Interarrival time	File (packet) size
Text	1 sec	mean = 500 bytes min. = 100 bytes max. = 10 Kbytes
Image	30 sec	mean = 120 Kbytes min. = 50 Kbytes max.= 150 Kbytes
Video	120 sec	mean = 50 Mbytes min.= 10 Mbytes max.= 120 Mbytes

In order to evaluate the performance of the IN Multimedia Video-on-demand service, three different simulation sets were considered. In the first set of simulations, we have been interested in evaluating the impact to the system performance of three major parameters of interest: the number of user interaction with the IP, the mean video file size and the mean average service time of the SSP. The results of this first set of simulations allows us to identify the main system parameters affecting the performance of the system. Five different cases are considered. In case 1, the basic value set applies. In case 2, we decreased the SSP mean service time to 3 ms. The parameters for case 3 are the same as the ones in case 2 except that the video mean file size is increased to 80 Mbytes. The parameters for case 4 are the same as the ones for case 2 except that the gateway service rate is increased to 30 Mbits /sec. The parameters for case 5 are the same as the ones in case 2 except that the video mean file size is reduced to 25 Mbytes. In all this first set of simulations, P_{MM} has been set to 0.10.

Figure 5 shows the call blocking probability for all five cases. For all five cases, the blocking probability increases as the number of user interactions increases. However, a closer look to Figure 5 shows that the mean video file is the major source affecting the system performance. From the figure, it is clear that when the mean file size has been set to 80 Mbytes (case 3), the blocking probability increases from 0.018 to about 0.024 as the number of user interactions increases from 2 to 6, an increase of 0.006 units, while for a mean file size of 25 Mbytes (case 5) the blocking probability increases only on 0.003 units for the same range of user interactions.

Figure 5 also shows that an increase on the processing power of the gateway alone does not help to improve the call blocking probability. This can be confirmed by the fact that in case 4 on which the gateway processing power has been enhanced, the increase on the blocking probability is slightly higher than 0.005 units. This clearly demonstrates that it is not only enough to increase the processing power of the gateway but the number of lines interconnecting both networks due to the heavy load imposed to the system when handling large video files.

The results for case 1 and case 4 also show that the gateway is the server prone to congestion. By comparing the results for case 1 to those when the service rate of the SSP increased (case 1) and those when the service rate of the gateway is increased (case 4), it is evident that the major source of blocking is the lack of memory space at the gateway. The impact on the blocking probability between the reference case (case 1) and case 4 is more important than between case 1 and case 2.

In the second set of simulations, we have been interested in analyzing the effect on the performance when transferring multimedia news containing video streams. In this set of simulations, the multimedia files being retrieved are characterized by the large amount of information to be transferred. The multimedia file sizes being considered are 2, 3, 5 and 10 Mbytes in size. We have fixed the number of user interactions between the IP and the user to 4. We also have assumed that the background traffic into the gateway consists solely of video files. As we have seen from the first set of simulation, video traffic is one of the major factors affecting the performance of the system. Other parameters are taken from the basic value set.

Figure 6 shows the call blocking probability versus the interarrival time of the background video files coming into the gateway from the Internet for the four different multimedia file sizes under study. Figure 6 shows that the multimedia file size has a larger impact on the performance of the system at high loads (shorter interarrival times of the background video traffic). As the interarrival time increases, the blocking probability for all the multimedia file sizes tend to the same value. This result may lead us to the conclusion that under the given system configuration, the size of the multimedia file has little or no impact over the performance of the system. However, a closer look to the result can provide a better insight on the effect of the different parameters over the overall system operation.

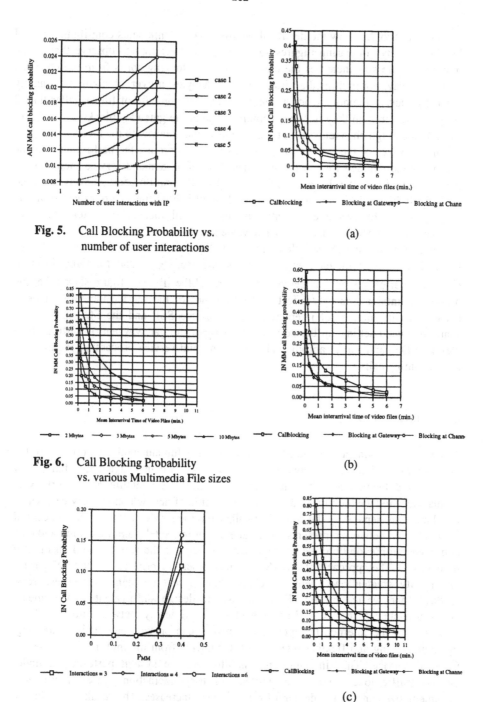

Fig. 5. Call Blocking Probability vs. number of user interactions

(a)

Fig. 6. Call Blocking Probability vs. various Multimedia File sizes

(b)

Fig. 8. Call Blocking Probability vs. P_{MM}

Fig. 7. Call Blocking vs File Sizes a) 2 Mbytes b) 3 Mbytes and c) 10 Mbytes

(c)

Figures 7a to 7c show the details of the two major components contributing to call blocking. By comparing Figures 7a to 7c, it is interesting to observe that as the multimedia file size increases the major source of blocking shifts from the channels to the gateway. For multimedia files of size of 3 Mbytes or over, blocking may mainly occur for lack of memory space at the gateways. For the case of files smaller than 3 Mbytes, even though they may be accommodated at the gateway, blocking will occur by lack of enough channels (lines) to interconnect the gateway to the PSTN.

The objective of the last set of simulations was to project the performance of the system as the demand for the IN multimedia News-on-demand service increases. In these simulations, we have assumed that the IN multimedia gateway is used exclusively by the IN multimedia service. The multimedia file size retrieved from the database server follows an exponential distribution with mean 400 Kbytes. All other parameters have been set using the basic value set.

Figure 8 shows the IN call blocking probability as a function of the probability of requesting the Multimedia News-on-demand service, P_{MM}, and for three different mean number of interactions between the user and the IP. The figure shows that the number of interactions may affect as the demand for the service increases. This is because as the number of interactions increases the SSP and the SCP have to be shared between setting up the connection between the gateway and the Multimedia News-on-Demand server and the call processing and service processing required by new incoming calls.

4 Conclusion

The telecommunications industry is undergoing a fundamental revolution with the introduction of intelligent networks. In this paper, we investigate this fairly new concept and technology. The main driving force of the intelligent network is to dramatically reduce the time and cost in the creation of new telephone services, which is achieved by separating the control of calls from the service control. The concept of the intelligent network represents a central element in the telecommunication and data communication evolution: the IN approach will become the way to build all kind of networks. The approach of this IN multimedia service brings up a new IN service opportunity which allows telephone subscribers to access abundant Internet resources. In this paper, we use existing IN technology to design and evaluate a Multimedia News-on-Demand service. With a Multimedia Gateway between the telephone network and the Internet, the intelligent network can provide automatic addressing, dynamic routing and flexible selecting of documents from the multimedia News-on-Demand service. The simulation results illustrate the system performance under various traffic load conditions, number of user interactions, mean multimedia document sizes and as the demand for the service increases. The results have shown that the stringent requirements when supporting video applications will require a careful planning of resources. The planning of the resources will involve not only the deployment of enough lines interconnecting different network platforms, but the use of specialized devices able to cope with the stringent processing requirements of new (multimedia) applications and intelligent services. Our results have also shown that the deployment of more sophisticated IN services will require a careful planning of the SCP resources.

Acknowledgments

This work was supported by a Bell Canada Research Contract. We are grateful to David Kraychick of Stentor, Fred Owen and Linda Amalu of Bell Canada for their help and cooperation.

References

1   Thörner, J *Intelligent Networks*, Artech House (1994)
2   Robrock, R B "Intelligent networking - narrowband to broadband", *Proc. Twelfth International Conference on Computer Communications,* Seoul (August 1995) pp. 189-194
3   ITU-T, Introduction to Intelligent Network Capability Set 1, 1993
4   ITU-T, Draft Recommendation, Introduction to Intelligent Network Capability Set 2, (1996)
5   Desmond, C and Shumada, I "Advanced Intelligent Network Tutorial" *Canadian Journal of Electrical and Computer Engineering*, vol. 19, no. 1, (1994) pp. 17-20
6   Hansson, A "Evolution of intelligent networks concepts" *Computer Communications* Vol. 18 No. 11 (Nov. 1995) pp. 793-801
7   Chong, H F and Wolf, C "Methodology and tools for intelligent network service specification", *Proc. IEEE Globecom'94,* San Francisco (Nov-Dec 1994) pp. 1243-1247
8   Yeh, S Y "Designing intelligent network architectures and services", *Proc. IEEE ELECTRO/94,* Boston (May 1994) pp. 271-275
9   Goerlinger, S "Intelligent network : The service creation environment" *Commut. Transmission* Vol 17, No. 2 (1995) pp. 13-20
10   Herman, M "Intelligent area wide Centrex" *Telcom Rep. Int.* Vol 18 (1995) pp. 34-38
11   Fujita, Y, Ookubo, K and Tokunaga, H. "Requirements for Intelligent Network Service Operations" *Journal of Network System Management*, Vol. 3, No. 2, (1995) pp. 195-216
12   Yan, J "Dimensioning network resources for IN services", *Computer Networks and ISDN Systems* Vol. 28 (1996), pp. 627-633
13   Söderberg, L "Evolving an intelligent architecture for personal telecommunication" *Ericcson Rev.* No. 4 (1993) pp 156-171
14   Steinmetz, R and Nahrstedt, K *Multimedia Computing, Communications and Applications,* Prentice-Hall International (1995)
15   PCIA, 1994 PCS Market Demand Forecast (January 1994) Washington, D. C.
16   Patel, S P, Henderson, G and Georganas, N D "The multimedia fax-MIME gateway" *IEEE Multimedia Magazine* Vol. 1 No. 4 (Winter 1994) pp. 64-70
17   Lamont, L., Lian, L, Brimont, R. and Georganas, N D " Synchronization of Multimedia Data for a Multimedia News-on-Demand Application" *IEEE Journal on Selected Areas in Communications*, Vol. 14, No. 1 January 1996, pp. 264-278
18   Pham, X H and Betts, R "Congestion control for intelligent networks", *Computer Networks and ISDN Systems* Vol . 26 (1994) pp. 511-524
19   MIL 3, Inc., "OPNET MODELER", Reference Manual, Release 2.4, 1993

Open Switching for ATM Networks

Manuel Duque-Antón[1], Ralf Günther[1], Raschid Karabek[2], Thomas Meuser[1], and Josef Wasel[1]

[1]PHILIPS Research Laboratories,
Weisshausstr. 2, D-52064 Aachen, Germany

[2]Technical University of Aachen, Dept. of Computer Science (Informatik IV)
Ahornstr. 55, D-52074 Aachen, Germany

Abstract. This contribution discusses the way how Open Switching can be supported in ATM networks. First the elements of an ATM system that are relevant for call control and call handling are described. Then we evaluate which of the call control interfaces should be realized by which parts of the system, especially which signalling aspects are relevant for the software architecture on network side. Finally we propose a concept for the integration of external call control as part of the network control architecture. The paper aims at a switch control architecture that facilitates for providing advanced applications in ATM networks by external call control. The definition of standardized protocols and the design of Open Switching applications themselves are out of the scope of our work.

1. Call Control in ATM Systems

The global architecture of an ATM system consists of a meshed network, which is built up by the interconnection of individual ATM switches and terminals that are attached as user or server equipment, see Fig. 1.

The terminal connections operate according to the well-defined *User-Network Interface* (UNI) and for switch interconnection the *Network-Network Interface* (NNI) is under discussion, see [1] and [2]. For the aspects considered in this paper we assume that the switches provide all control functions according to the standardized signalling protocols. On the user side all ATM-relevant functions, including the user side signalling, are performed by the ATM adapter card and the associated software driver.

The ATM protocol reference model distinguishes between a user plane to transport user information and a control plane dealing with signalling information providing *out-of-band signalling*. Signalling is performed between the end-stations using a message based protocol over a separate channel provided by the network [3]. A call is initiated by the calling user party which sends a call request via UNI into the network, i.e. to the switch where it is connected to. Each switch that receives a call request via UNI or NNI processes a *Switch Signalling* procedure. Thereby it interacts with the internal switch control, in order to perform its local call admission and resource allocation procedures. If enough resources are available to admit the call, the request is forwarded to the next switch via NNI or - if the addressed destination is directly connected to the switch - the call request is send to it via UNI. If the call has to be refused a call reject is sent back to the calling party. Fig. 2 shows the signalling protocol stacks on user and network side.

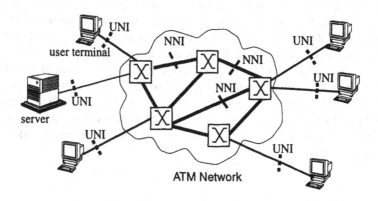

Fig. 1: Global architecture of an ATM system

The Open Switching architecture is a versatile concept for external call control by applications [4] which requires for access to Switch Signalling. Thus, in order to allow for Open Switching we have to consider the structure of the corresponding switch architecture. The rest of the paper is organized as follows: First, we introduce a software architecture for control and management of ATM switches. After a short explanation of the basic concepts for Open Switching, several Open Switching applications in ATM systems are presented, and a global Open Switching architecture is proposed.

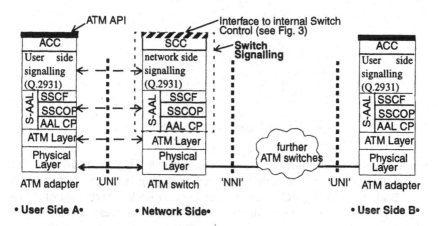

Fig. 2: Protocol stacks for control plane on user and network side, where ACC corresponds to Application Call Control, SCC to Switch Call Control, SSCF to Signalling Specific Coordination Function, SSCOP to Service Specific Connection Oriented Protocol, and AAL CP to AAL Common Part

2. The Switch Architecture

As part of a software architecture [5], that we have developed for ATM switches, a switch control software architecture has been defined. The proposed architecture is intended to cope, basically, with distributed ATM switches as described in [5], but is also able to handle centralize versions of it. Is major properties are a distribution platform for a location-transparent implementation and operation of object-oriented *Switch Control Applications* (SCAs) and a principle for structuring of switch control software independently of the underlying physical configuration. For the latter aspect a *Switch Control Interface* (SCI) is introduced that allows the use of the switch-internal control block by independent SCAs, see Fig. 3. This interface provides service primitives for set-up, modify and release connections as well as for configuration, routing and capacity control.

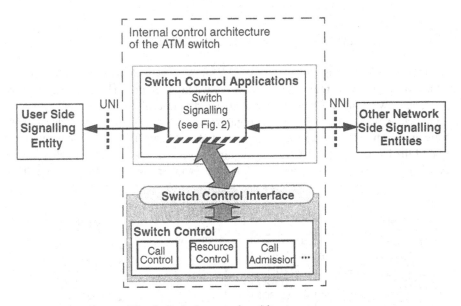

Fig. 3: Switch control architecture

Switch Control Applications may run on every switch. They are realized as software objects that interact to each other and to the switching system via a distribution platform. Similar to CORBA (Common Object Request Broker Architecture) [6] the employed distribution platform provides a software bus for the interoperability of distributed objects in heterogeneous environments. This includes addressing of objects, setting up communication links, and invoking operations on objects. The distribution platform is running on several operating systems (*Windows NT, Solaris, pSoS*). Thus the integration of general purpose computer systems with the ATM communication devices can be realized.

CORBA and the distribution platform used here have similar functionality and both are operating according to the same principles. Our target usage environment, however, which aims to distributed real-time control application requires for a dedicated solution. Thus we decided not to use a CORBA-compliant ORB but a more real-time, light-weight ORB. Our distribution platform supports multi-tasking, asynchronous procedure calls and multi-priority queuing. It allows for a direct use of ATM real-time communication facilities and addressing information. We restrict our platform only to those services required in our specific usage environment. Services for naming, persistence, concurrency control, and software exchange are installed as a inherent part of the ORB instead of being Object Services in the sense of CORBA. This choice, however, does not imply any restrictions to the installation of our ATM switch in a standardized way. The distribution platform is used only as base of the switch control architecture. The interactions to external user and server terminals follows the standardized ATM signalling and - as we will discuss latter - a standardized Open Switching protocol.

Regarding the concept of Open Switching in ATM networks that is discussed in this paper it is not relevant which distribution platform is employed. But it is necessary that SCAs can be designed and implemented independently of other control tasks and the internal switching system. Furthermore the means for the interaction of software objects have to be provided.

For switch internal processing SCAs interact with Switch Control objects (e.g. Call Control or Call Admission Control) via the above-mentioned SCI. For external communication they use a specific protocol via a well-defined VPI/VCI channel, e.g. signalling is done according to the UNI/NNI protocol and management applications use SNMP. The functional component "Switch Signalling" shown in Fig. 2 respectively Fig. 3 is installed as one of several SCAs. After having received a signalling packet via UNI or NNI, the switch signalling maps it to the appropriate signalling instance. Then it translates the received primitive into a corresponding SCI message by which it invokes appropriate actions from Switch Control objects.

Our architecture allows for installing any Service Switching Function (SSF) as SCA beside other Switch Control objects. These SSFs can interact with the signalling objects and the internal switch control via the distribution platform. For the interaction to external Service Control Points (SCPs) they use any message-based Open Switching interface via pre-established ATM channels.

3. Open Switching Architecture

Before showing the concept of embedding the SSFs into the switch control architecture we summarize the relevant aspects of Open Switching. The Open Switching architecture is a versatile concept for external call control by applications [4]. The motivation for such an architecture is the ever increasing demand for the fast and flexible introduction of new applications and services[1] in telecommunication networks.

---

1. The words 'service' and 'application' are deliberately used as synonyms although there are subtle differences between these two.

We can identify three general requirements for such kind of telecommunication applications:

1. Flexibility to create new services and to modify parameters of existing ones.

2. Processing information that is stored in general purpose databases.

3. Call control by applications, i.e. the application gets information about the call status and may influence call handling by requesting additional services.

Having in mind the service control architecture from Fig. 3 it seems natural to support a concept that distinguishes between the "Basic Switching Function" and the "Application Provision Function". Running the applications on a conventional computer platform means that services and support for them can be created and managed efficiently by available data-processing technology. The basic switching function is responsible for build-up, maintenance and release of connections but has no knowledge about service logic. The two components – basic switching function and application provision function – communicate via message-based interfaces.

Thus Open Switching requires well-defined interfaces and an explicit description of how call handling is structured. For the latter purpose the *Basic Call State Models* (BCSM) has been introduced. The BCSMs are based on ITU-T's IN recommendations and define call phases, detection points and transitions between both.

The Open Switching Reference Architecture [4] shown in Fig. 4 defines interfaces that separates the Basic Switching Function (BSF) from the Application Provision Function (APF). A message interface is offered between both components. The BSF is further divided in:

- The Call Control Function (CCF) which builds-up, maintains and releases connections. Detection Point (DP) Processing contains mechanisms to detect service requests and to process incoming control messages from the application. Both are extensions to the existing call handling software. The CCF performs without any information about applications.

- The Service Switching Function (SSF) which offers a logical view on call handling that abstracts from service-specific details. It also defines the message interfaces to the Application Provision Function.

During the call phases, usual call handling takes place: Signals from one of the two call parties are received and processed. From the point of view of the SSF, the call phases are atomic – i.e. call handling is not interruptible – therefore these phases are called Points In Call (PIC). At the Detection Points (DP) it is checked whether an event report has to be sent to the SSF. Only at the DP messages from the application control can be received and processed.

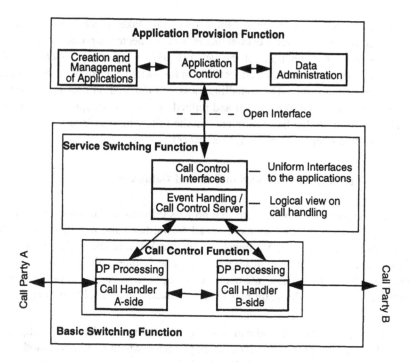

Fig. 4: Main parts of the Open Switching Reference Architecture

The event reports inform the applications about the currently reached DP. The applications respond either by sending a message to let call handling continue or – if an application has requested to take over call control – a directive is given to change the call context. The separation of call handling into PICs and DPs is the basis for a call state model that enables the application control to abstract from switch-internal details. The services only need an abstract description of what happens during the PICs and the information which DPs are supported. It depends on the granularity of the call state model which applications can be supported.

The Application Control executes applications that control call handling. It is responsible for presenting the network resources, the control of call handling and the message transfer to the switch. The interface between the Application Control and the Call Control Interface of the SSF can be designed according to each protocol supported by Open Switching, e.g. INAP [7] and ECMA CSTA [8].

4. Open Switching in ATM Systems

In the existing telecommunication switching networks most of the service and switching logic reside in (or are connected to) centralized components (exchanges) while the customer premises equipment is relatively simple structured. For ATM systems we have a totally different situation. Switching takes place between powerful end-user equipment, i.e. user workstations/PCs or connected servers, using only signalling and switching functions of the network. Following today's ATM signalling standards it is not foreseen to introduce DP's in the switch control part, that processes signalling on network side. The signalling messages initiated on user side received at the ATM network are coarse-grained and replied only by an accept or an reject message. Having in mind this basic difference in the following we discuss the realisation of Open Switching applications in ATM networks.

4.1 Processing Information stored in External Databases

Typical examples for applications that process information stored in external databases are personal communications (e.g. call forwarding) and intelligent networking (virtual private networks (VPN), freephone services). Using terminals that have only minimum intelligence, the application at user side is only able to produce dialling information. Thus, the earliest interaction with application control is performed after network access leaving the main part of the necessary functionality on the network-side.

In ATM systems the terminal equipment of the calling user has the intelligence to produce the dialling information after having decided which end point (another user, an IP server or other member of a VPN) should be called. Then the signalling entity on the adapter card is able to initiate a call set-up directly with the appropriate address information. Thus, the earliest interaction with application control can be performed before network access leaving the main part of the application control to the user-side.

As an example let us consider the set-up message in a VPN. Following the traditional approach the switch control receives an ATM set-up message. Having recognized that this message is a VPN service request the switch control would have to ask again the user for a short number. For the translation of this short number to a complete ATM destination address, again an the Application Control has to be called. In this scenario it seems a more natural way to introduce the VPN aspects at the user-side signalling where the VPN information is directly processed and translated into a basic ATM call request that includes a complete ATM address.

Contrary to that the Open Switching application *Call Forwarding* is an example that it is still useful to perform the earliest interaction with application control after network access leaving the main part of the necessary functionality on the network-side.

Fig. 5 shows the scenario for a connection set-up with call forwarding. User X and Y are both connected to an ATM system. User Y has moved from location B to location C. Now the connection set-up can be described by the following steps:

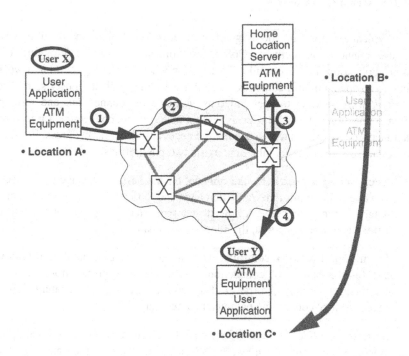

Fig. 5: Call Forwarding scenario

1. User X residing at location A initiates a connection to user Y by sending a set-up message to the switch it is connected to. That message includes an ATM destination address that indicates the called user (user Y) and also the switch it usually is connected to (location B).

2. The ATM system forwards the set-up message to the switch at location B.

3. The Service Switching Point of that switch recognizes that user Y is not reachable at this location. Therefore it asks the corresponding Application Control, a connected Home Location Server for example, for the actual address of user Y.

4. Using this address information, the set-up message is forwarded to the actual location of user Y.

For the call forwarding application it seems advantageous that the Application Control is not installed on user-side, because in that case the actual location information on each user has to be available at each host. Please note that in the discussed scenario we do not consider the different approaches to realize location control inside an ATM network.

4.2 External Call Control

Typical examples for the usage of external call control are the handling of bundles of user connections, including QoS negotiation and monitoring, belonging to a single user application. Let us consider the situation where a user logged in at a workstation intents to open a multimedia session. In this case, the application uses some - so called - *middleware* services to open video as well as audio connections and some possible data connections for file transfer. A session management cooperates, for instance, with a transport network control to acquire the necessary network resources. For the sake of accessing the network, mainly there exists two possibilities:

- After having triggered by the application the session management specifies the maximum and minimum traffic demands and QoS parameters for each connection. After the transport network control has requested for appropriate connections, it left the network alone with the admission decision.

- The transport network control starts with some traffic demands and QoS parameters for the bundle of connections. If the network gives information on some resource restrictions, the transport network control begins an interaction with the switch control concerning parameter re-negotiation.

If we assume that only the ordinary UNI interfaces exists, the second approach is not possible. The transport network control has to translate the handling of a whole bundle of connections into a sequence of UNI messages. Now the switch signalling has no information about the correlations between the requested connections. Thus, there is not any external call control and the application control can only be done at the user side. Of course, this is not very efficient: If some connections of the whole bundle can not be accommodated in the network, the transport network control is informed via the corresponding UNI signalling message (call reject). Thus probably, the session control asks the transport network control to release the already established connections of the bundle also.

If we can assume that there is an extended UNI or an additional control interface beside the standardized UNI, the transport network control should be able to deal with bundles of connections in a sophisticated way. Either the UNI directly supports the request of a bundle of connections or the transport network control can get detailed information about the switch resources and the state of its call request so that it can coordinate the individual calls for the connections belonging together. If we extend the UNI to accept calls for bundles of connections, the coordination of the single call establishment is burden to the switch control application. Otherwise this coordination remains by the transport network control on user side and the switch control application has to provide all necessary call state information.

Regarding the switch control architecture on network side we have to check whether the Switch Control Interface is prepared to cope with the requirements of such new interfaces. In general, the SCI promises a generic approach to access to switch internal control by Switch Control Applications. Thus if the required functions for external call control will become clear, we have to design a SCA and appropriate SCI primitive have to be defined.

Regardless the adopted approach, however, the Application Control remains outside the switch. The Switch Control only offers access to the needed information via the Switch Control Interface, in order to enable the Application Control to make the right decisions. Thus external call control in ATM networks can be achieved as a joint operation of user and network sided control functions.

5. Integration of External Call Control

5.1 Requirements for Open Switching on ATM networks

Having in mind the examples discussed above, the network part of an ATM control architecture has to support intelligent and interactive call control. This means, instead of requesting individual connections via UNI without any interaction between switch and terminal, establishment of correlated connections can be done step by step getting informations about the call state and the available resources. This implies the following requirements on the ATM signalling:

- The processing of detection point into the call process has to be introduced which allows for influencing the call during processing and for getting information about the state of the call request.

- Switch state information such as available resources, call admission procedures has to be made accessible.

- A communication connection between the SSF and the Application Provision Function either as extension of the UNI or as separate pre-established communication channel has to be realized.

- The translation of call control interface primitives to switch internal commands has to be supported.

- A coordination function to handle the DP events during signalling processing and commands coming from user side has to be supported.

5.2 Global Open Switching Architecture

Our approach for the design of an external call control interface is based on the integration of the aforementioned Open Switching functions into the existing ATM control architecture. The resulting architecture is shown in Fig. 6.

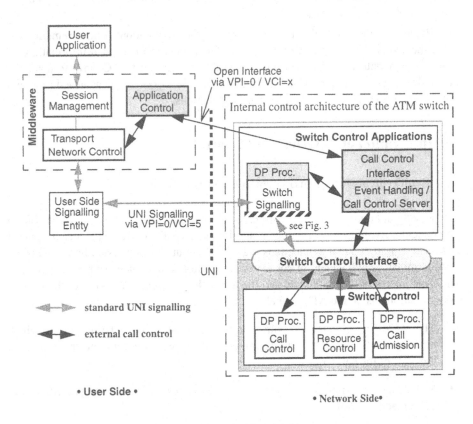

Fig. 6: Global Open Switching Architecture

Following the design and implementation rules specified in the software architecture the SSFs of the Open Switching concept can be realized as Switch Control Application. The *Event Handler, Call Control Server*, and the *Call Control Interface* are implemented as independent objects interacting via the distribution platform. Access to the switching functions is realized by the Switch Control Interface. This allows for an easy integration of the required coordination and translation functionality by just adding some new embedded software objects.

It is for further study whether the DPs should be defined within the Switch Control (Call Admission Control, Call Handler, Resource Control) or the Switch Signalling. In the first case the SCI have to be extended for supporting transfer of event reports from the DP Processing and control messages to the DP Processing. The second approach keeps the SCI unchanged and the interaction between DP Processing and Event Handling/Call Control Server is based on inter-object-communication. Therefore the state machine of the existing switch signalling, that allows only for one detection point yet, has to be extended. It is important to know, that our architecture supports both kinds of detection points. It has to be discussed which DPs are needed and how to standardize them.

Our architecture allows for two alternatives to realize the interaction between the control entities on user and network side. Using the approach as shown in Fig. 6 there are *separate ATM channels* for UNI and the Open Interface. All applications running on top of the middleware give their communication requirements (e.g. a bundle of connections each with QoS parameters in a specified range) to its associated session management. The Transport Network Control is aware of the correlations between connections belonging to the same application. Using the standardized UNI for switch processing it requests the individual connections one after another.

By processing the Detection Points that have been introduced into switch signalling and switch control the progress of the connection set-ups is controlled. The Event Handler coordinates the actions to be done in response to the information provided at the DPs. If necessary the Event Handler informs the Transport Network Control about the state of its connection requests.

If the Transport Network Control receives a notification or control message via the Open Interface, it interacts with an Session Management in order to decide the progress of the call. Maybe the call parameters will be changed or the call is cancelled. The requirements of the connections belonging to the same bundle can be modified in order to support the whole application at least with basic functionality. If the minimal requirements can not be fulfilled, the application is cancelled. The appropriate order is send back to the switch via the Open Interface. Finally the User Side Signalling entity is informed only about the success of its request via the standardized UNI as before. This approach leaves the ATM signalling unchanged.

As alternative solution we can combine the UNI and the Open Interface. Setup as well as all interaction during call handling are done via an extended UNI. Therefore the standard UNI has to be changed. In this scenario switch signalling is not called by the user side signalling directly. All interaction is done via the Call Control Interface which initiates a call setup to the switch signalling internally.

6. Future Work

It is for further discussion which Open Switching applications make sense for ATM networks and which protocols should be supported at the call control interfaces. The standardized protocols like INAP or CSTA came into play with the emerging activities in the area of *Voice over ATM*. Wireless ATM systems require for user-mobility services. Probably additional protocols have to be defined for new, more-demanding applications like multimedia conferencing. In any case international standards have to be supported by the call control interface within the Open Switching architecture.

It is also for further study whether the DPs should be defined within the Switch Control, i.e. in the Call Admission Control, the Call Handler, and the Resource Control, or the ATM Switch Signalling. In the latter case the *ATM Forum* as standardisation organisation has to extend the existing UNI signalling protocol by a call state model that supports the required detection point mechanisms and enable call manipulation by applications. It seems to be a more pragmatic way to realise external call by running the Open Switching application on a separate channel leaving the ATM signalling unchanged.

For supporting of multimedia and conferencing services the discussed Open Switching concepts have to be extended by models for the session management and transport network control. Related work on architectures for middleware services is under way in many research activities around the world.

References

[1]   ATM Forum/94-1018R7: *User-Network Interface, Signalling, Version 4.0*, October 1995.

[2]   ATM Forum/af-pnni-0055.000: *Private Network-Network Interface, Specification Version 1.0*, March 1996.

[3]   B. Law "Signalling in the ATM network", BT Technol J., vol. 12, no. 3, July 1994.

[4]   M. Elixmann, B. de Greef, A. Lelkens, K. Neunast, H. Tjabben "Open Switching - Extending Control Architectures to Facilitate Applications", XV International Switching Symposium ISS'95, vol.2, April 1995, pp. 239 - 243.

[5]   M. Duque-Anton, R. Günther, R. Karabek, T. Meuser: *A Software Architecture for Control and Management of Distributed ATM switching*, Proceedings of KiVS'97, Braunschweig, in Informatik Aktuell, Springer Verlag, February 1997.

[6]   Object Management Group OMG: *The Common Object Request Broker: Architecture and Specification* Revision 2.0, July 1995.

[7]   CCITT Study Group XI "Intelligent Network Series Q.1200".

[8]   ECMA "Services and Protocols for Computer-Supported Telecommunication Applications (CSTA)" ECMA, June 1992.

Application of INAP to AIN Intelligent Peripheral

Hee Jin Lim, Go Bong Choi

AIN Switching Section
Electronics and Telecommunications Research Institute (ETRI)
P.O.Box 106 Yusong Taejon 305-600, KOREA
E-mail: hjlim@etri.re.kr

Abstract. As the infrastructure of the advanced intelligent network (AIN) has evolved, the intelligent peripheral (IP) has emerged as a major node for adding value to telephony services. It provides user friendly interfaces for customizing a service with various resources such as DTMF digit collection, voice announcement, voice recognition, video payload, and audio conference bridge, etc. We are developing a kind of networked IP which supports ISDN user part (ISUP) and intelligent network application protocol (INAP) for the communication with the service switching point (SSP) and the service control point (SCP) respectively. This paper describes the AIN IP architecture, telecommunication interface subsystem which is one of the four subsystems of IP, and functional blocks of INAP. The IP INAP provides application services by collating with the SCP, processed messages, and operations carried out at the IP, to provide specialized resource services with 3 ASEs which provides 7 operations. We applied this system to interface scenarios between the IP and other network elements and user interaction procedures.

1 Introduction

The advanced intelligent network (AIN) is a service independent architecture in which new services are offered easily and effectvely using service independent building blocks (SIB). It is a structural network concept applicable to all networks, such as PSTN, ISDN and PLMN, providing platforms capable of easily creating services according to user orders in multi-vendor environments[1].

Intelligent peripherals, major nodes of AIN, contain centralized resources difficult to be maintained at a service switching point (SSP) or a service control point (SCP), and provide high-quality services to intelligent network users operating a variety of media and interfaces, when so requested by an SCP [2,3,4].

The AIN IP provides specialized resources such as play announcements, DTMF collection, prompt and receive/synthesize message, fundamental level message recognition, and fax processing functions, during the initial stage of development, with plans to gradually expand the range of services.

The AIN IP system is designed with 4 major subsystems: a specialized resource control subsystem (SRCS) which controls boards used to provide various resources (voice, links, etc.); a telecommunication interface subsystem (TIS), which communicates with external network nodes and transfers signal messages; a specialized resource functions (SRF) service control subsystem (SSCS), which drives related resources with service processing logic within an IP and manages the conditions of a variety of resources; and a system operation and management subsystem (OMS), which operates and maintains the system .

This paper describes the structure of the network interface part which interfaces with the SCP and the SSP, paying particular attention to INAP's functional blocks, message design, and processing operations, to implement AIN IP on the basis of INAP[5,6] recommended by international telecommunication union-telecommunication standardization sector (ITU-T) for communication between intelligent nodes.

2 AIN IP Overview

2.1 AIN IP Architecture

The AIN IP communicates with the SCP directly as shown in Figure 1. The communication method between the IP and the SCP is ITU-T CS-1$^+$ INAP which is the extended CS-1 INAP for message processing services. If a local IP has no resource for a service, the SSP has accessibility to a remote IP via an assisting SSP or local exchange through ISUP. By 1998, we will be deploying four AIN services, that use IP efficiently, such as advanced freephone, advanced credit calling, televoting, and wired universal personal telecommunication (UPT) in the first phase.

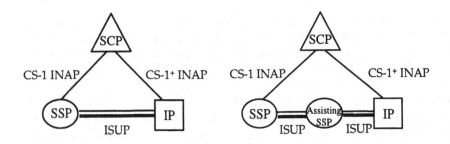

Fig. 1. AIN physical node interface

As shown in Figure 2, the physical structure of an AIN IP consists of a main processor (MP), a specialized resource switching processor (SRSP), and a specialized resource control processor (SRCP), and the 4 aforementioned subsystems are distributedly installed.

Some of the functions of the SRF service control subsystem, and system operation and management subsystem, are installed at MP. The high-quality specialized resource control functions, such as the prompt and receive message/ message recognition and synthesis, and fax processing, are installed at SRCP. The SRSP is located at the front end of SRCP, where some of the functions of the telecommunication interface subsystem and SRF service control subsystem are installed, to interface with SSP and SCP. On the other hand, MP and SRSP are connected to Ethernet; MP and SRCP to LAN; and SRSP and SRCP are connected to E1 links. This structure guarantees a scaleability which easily allows the extension of the AIN IP service processing capacity by adding new SRCPs, according to service accommodation capacity.

We reduced the Ethernet traffic and MP loading, by enabling the system to provide basic services, such as DTMF collection or playing announcements, with only SRSP installed at the front end, and promoted economic efficiency by improving service provision speed, and by reducing the frequency of expensive SRCP usage. If a user wants to use some special resources such as voice recognition, voice recording, and fax sending/receiving, the SRCP is connected to the SRSP under the control of the SRF service control subsystem located at the MP.

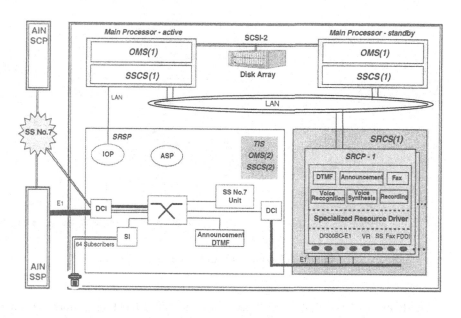

ASP	Access Switching Processor	IOP	remote I/O Processor
DCI	Digital CEPT1 Interface	SI	Subscriber Interface

Fig. 2. AIN IP architecture

2.2 Telecommunication Interface Subsystem (TIS)

The AIN IP telecommunication interface subsystem supports an ISUP protocol for an SSP interface, and an INAP for an SCP interface. In other words, the system exchanges control signals with an SSP and an SCP through SS No.7 signal networks, transmits voice and data information to users via an SSP through an E1 trunk, and processes a variety of control messages with respect to errors and alarms occurring on signal networks, ensuring satisfactory communications.

All functions of the TIS are located on the SRSP, and are designed as a hierarchical structure by matching with SS No.7 protocol levels. Furthermore, individual functions within the application protocol stack have been modularized, to minimize correlations when interacting with the SRF service control subsystem, and to facilitate the addition of new operation processing functions.

As shown in Figure 3, the TIS is hierarchically designed and delivers control signaling and bearer information (voice, text, picture) through the SS No.7 signaling network and E1 trunk respectively.

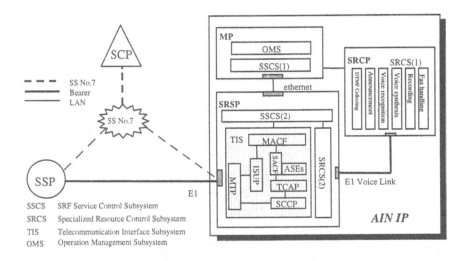

Fig. 3. Telecommunication interface subsystem

Our IP applies CS-1$^+$ INAP which partly includes CS-2 INAP operations. To support the ISDN user, the IP communicates with the SSP by ISUP. Also, the ISUP should be enhanced to transfer AIN service-related information, such as SCP id and correlation id, as well as to provide setup/release of connection procedures. Transaction capabilities application part (TCAP) which was implemented for pre-IN, should be modified to include ITU-T white version recommendation for supporting the AIN INAP. Other protocol stacks such as SCCP and MTP will be applied like pre-IN's.

3 IP INAP Design

3.1 Functional Blocks

INAP, composed of a multiple association control function (MACF), a single association control function (SACF), and application service elements (ASEs), is a protocol which processes application services by exchanging control signals with an SCP via a signal network through TCAP, signaling connection control part (SCCP), and message transfer part (MTP).

We designed 4 functional blocks as IP INAP component blocks – a protocol message matching (PMM) block, corresponding to MACF, and 3 functional blocks corresponding to SACF, i.e., single association control management (SACM), single association operation control (SAOC), and single association dialogue control (SADC).

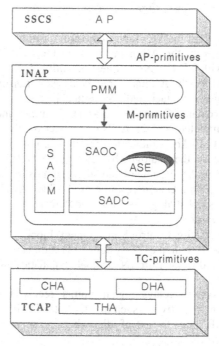

Fig. 4. IP INAP functional block structure

- PMM (Protocol Message Matching) block
  This block interfaces with the application process (AP) of the SRF service control subsystem, and with the telecommunication interface subsystem. In other words, the

block analyzes messages received from the ISUP or the SACF, transmits appropriate service primitives to the AP, and transmits primitives received from the AP to the SACF and ISUP.

In addition, if the block receives information regarding signal point congestion, it adjusts traffic according to control algorithms. If time out occurs during connection to the SSP and user interaction with the SCP, controlling the Tsrf timer sends a signal to the AP, ISUP, and SACF to release resources to the idle state.

• SAOC (Single Association Operation Control) block

This block exchanges M-primitives related to operation requests/indications with the PMM, and exchanges TC-primitives with the component handling functional block (CHA) of the TCAP. SAOC activates applicable ASE according to the kind of message received, to carry out operation encoding or decoding; sequentially executes operation instructions received from the SCP, by maintaining operation sequence. Also, it provides error checking functions with respect to components and operations received from the SCP.

• SADC (Single Association Dialogue Control) block

This block interacts with the SAOC, and allocates, maintains, and releases dialogue numbers for each service call, and carries out application context negotiation (ACN) functions, by interacting with the TCAP. The ACN procedure, which guarantees satisfactory communications between physical entities in mutually different evolutionary stages, as follows: First, the IP sends an AC name after putting in TC-Begin request primitive.

SCP INAP then receives the TC-Begin indication primitive; and, if the AC name is found to be different from the AC name held by the SCP INAP, the AC name is substituted and sent with TC-U-Abort request primitive. If the application context can be provided as SCP ASEs, the AC name, received when transmitting the TC-Continue request primitive, is inserted intact, and transmitted as it is.

When the IP receives an AC name such as TC-Continue indication primitives, the dialogue setting is completed, and the ACN ends successfully. If the IP, after sending TC-Begin, receives TC-U-Abort indication primitives, it verifies whether a substitute AC name is there, restarts the dialogue by sending a new TC-Begin request, if the IP INAP can accommodate the AC name, and sends signals to the AP, so that the ASE, equivalent to substitute the AC name, can be used. Such ACN processes are required only during the process of dialogue setting. Once the dialogue is set, the AC name exchanges are not necessary.

• SACM (Single Association Control Management) block

This block interfaces with the IP system operation and management functions. When messages required to measure INAP-related statistics and data occur, a advises of such an occurrence.

In addition, the system relays signal points received from the TCAP, and subsystem management primitives (such as TC-State, TC-Pcstate, etc.), to the PMM, and converts the INAP status to that of idle, when the service is stopped for reasons such as signal point error.

Table 1 defines TC-primitives used between TCAP and SAOC, and SADC. Also, M-primitives are used between PMM and SACM, and SAOC, within INAP. AP-primitives including bearer channel related messages used for communications with SSP through ISUP, along with messages defined as M-primitive, are not discussed here.

Table 1. INAP-related messages

PRIMITIVE	SEND	RECEIVE
TC-primitives	INP_tc_beg_req INP_tc_con_req INP_tc_end_req INP_tc_u_abt_req INP_tc_inv_req INP_tc_res_req INP_tc_u_err_req INP_tc_u_rej_req INP_tc_state_req TCAP_pid_req	INP_tc_con_ind INP_tc_end_ind INP_tc_u_abt_ind INP_tc_p_abt_ind INP_tc_not_ind INP_tc_inv_ind INP_tc_u_err_ind INP_tc_l_can_ind INP_tc_u_rej_ind INP_tc_l_rej_ind INP_tc_r_rej_ind INP_tc_state_ind INP_tc_pcstate_ind TCAP_rest_ind TCAP_pid_ind
M-primitives	INP_pcui_to_macf INP_pa_to_macf INP_prm_to_macf INP_can_to_macf INP_rej_ind INP_dial_rel_ind INP_pid_ind INP_Tsrf_reset INP_rest_ind INP_state_to_macf INP_pcstate_to_macf	INP_ari_needed INP_pcui_resp INP_pa_resp INP_prm_resp INP_can_resp INP_dial_abort INP_pid_req INP_macf_state_req INP_Tsrf_expired

3.2 Operation and Error Processing

ITU-T CS-1 recommends the provision of 59 operations, with a total of 25 ASEs, to define INAP, the communications protocol between intelligent network physical entities. Korea Telecom (KT)-INAP[7], in order to gradually evolve to advanced intelligent network, has established standards to provide a total of 35 operations, with 22 ASEs, to meet domestic circumstances. Of these operations, the types of operation used between the AIN IP and the SCP are related to service start and cancellation, and to user interaction for utilization of specialized resources. The AIN IP system provides a total of 7 such operations, with 3 ASEs, as shown below, thus providing all functions, except for 4 script-related operations among the 11 operations related to the IP of CS-2 INAP.

- SCF/SRF activation of assist ASE: assist request instruction (ARI) operation
- Specialized resource control ASE: prompt and collect user information (PCUI), play announcement (PA), prompt and receive message (PRM), erase message (EM), specialized resource report (SRR) operation
- Cancel ASE: Cancel operation

Table 2 defines the parameters of each operation, and the timers required to execute operations.

Table 2. IP INAP processing operation

Operation		Timer		Detailed Parameters	Remarks
Title	Grade	Limit Value	Node		
ARI	2	Tari (2 seconds)	IP	CorrelationID	
PA	1	Tpa (30 seconds)	SCP	InformaionToSend DisconnectFromIPForbidden RequestAnnouncementComplete Extension	■ Includes fax sending and voice synthesis function
PCUI	1	Tpcui (65 seconds)	SCP	CollectedInfo DisconnectFromIPForbidden InformaionToSend Extension	■ Includes voice recognition function
SRR	4	Tsrr (1 second)	IP	Null	
Cancel	2	Tcan (3 seconds)	SCP	InvokeID or All Requests	
PRM	1	Tprm (65 seconds)	SCP	DisconnectFromIPForbidden InformaionToSend SubscriberID MailBoxID InfoToRecord Extension	■ Includes fax receipt function ■ CS-2 operation
EM	2	Tem (3 seconds)	SCP	DisconnectFromIPForbidden MessageIDtoErase CallSegmentID Extension	■ Recorded message erasure function ■ CS-2 operation

As shown in the table, the timer actually operates at the TCAP of a side (i.e. the node where TC-Invoke request primitives have occurred) which has transmitted operation execution requests. If time out occurs, it sends signals to the INAP of its own station, by using TC-L-cancel primitives to take follow-up action. The timer values are loaded from the real time database of the IP. The values can be modified through the authorized access. The message recognition function of this AIN IP system is driven by the detailed parameter value of PCUI operation, while the message synthesis and fax sending functions are provided by the PA operation without a new operation definition.

Table 3. Error codes for IP

Operation \ Location	INAP	AP
PCUI	7,8,14,15,16	0,4,11,12,13
PA	7,8,14,15,16	0,11,12,13
Cancel	7	1,12
PRM	7,8,14,15,16	0,4,11,12,13,19
ARI	7,14,15,16	6,12
EM	7,8,14,15	11,12,13,20
SRR	*	*

Code	Description	Code	Description
0	Canceled	12	Task refused
1	Cancel failed	13	Unavailable resource
4	Improper caller response	14	Unexpected component sequence
6	Missing customer record	15	Unexpected data value code
7	Missing parameter	16	Unexpected parameter code
8	Parameter out of range	19	Unknown subscriber
11	System failure	20	Unknown recorded message ID

Furthermore, the PRM of CS-2, and the EM operation, have been introduced early for prompting and receiving message, and fax receiving services. Meanwhile, errors that may occur in the IP INAP's individual operations fall into 14 categories as listed in Table 3.

As SRR is a grade-4 operation, error defining is not required and ARI errors are detected at the SCP INAP side. The errors on other operations occur at the INAP of the AIN IP, or at AP of the SRF service control subsystem. We grouped the errors into the AP detected and the INAP detected.

4 Interface Scenarios

This chapter describes the connection setting and interaction between the IP, the SSP and the SCP centered in the AIN IP network interface part (SCCP and MTP omitted), and release procedures, along with ACN. This scenario applies, in common, to all kinds of intelligent network services utilizing IP.

- Connection Setup

Figure 5 shows IP service starting procedures. If, during AIN service progress, a situation requiring the use of specialized resources occurs, the SCP issues instructions to the SSP to connect with the IP. If the IP receives a request for bearer connection from the SSP, it allocates available channel, and sends a TC-Begin request primitive to the SCP to set up dialogue, along with ARI operation. At this time, an AC name is loaded onto a primitive transferred to TCAP. The AC name is R2 and C1, i.e., R2 (Provides SCF/SRF activation of assist ASE and specialized resource control ASE) and C1 (Provides Cancel ASE), representing the range of operations that can be provided from IP INAP.

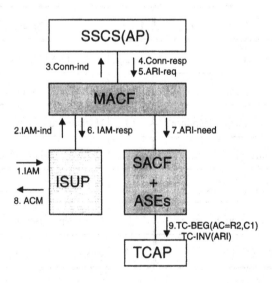

Fig. 5. Connection setting and dialogue start procedures

- User Interaction

Figure 6 shows a situation in which the ACN has succeeded, and SCP has begun user interaction procedures. In other words, IP INAP receives a TC-Continue primitive and completes the dialogue setting, and processes services after receiving detailed operation execution instructions.

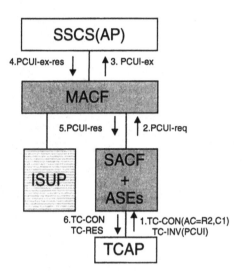

Fig. 6. User interaction procedures

- Disconnection

Disconnection procedures consist of IP initiated disconnection and SSP initiated disconnection. IP attempts to disconnect in accordance with the parameter DisconnectFromIPForbidden values, among the operation parameters transmitted. Meanwhile, when the Tsrf timer expire is sent from the PMM, all pending components are discarded, and TC-U-Abort is transmitted to stop communications with the SCP.

Figure 7 shows the IP initiated disconnection procedures which consists of pre-arranged end and basic end. Pre-arranged end means that the TCAP of the IP doesn't need to send a message for the dialogue end because the SCP releases dialogue according to its own service logic.

Figure 7 (a) shows an example a pre-arranged end. If the SCP sends a PA operation in which a requestAnnouncementComplete item is false and the disconnectFromIPForbidden is false, after successful PA processing, the IP attempts to disconnect from the SSP and release dialogue with the SCP. Figure 7 (b) shows an example basic end. When the SCP sends a PCUI operation in which the disconnectFromIPForbidden is false, the IP sends TC-End primitives set to basic end with successful results to the TCAP. Then the TCAP of the IP transmits the transaction message to the SCP to end the dialogue.

In Figure 8, SSP initiated disconnection procedures are shown. In this case, the SCP sends disconnectForwardConnection operation to the SSP and then releases dialogue between the IPs. When an IP receives the release message from the SSP, the IP releases the resources of the call and ends the dialogue by pre-arranged end procedure.

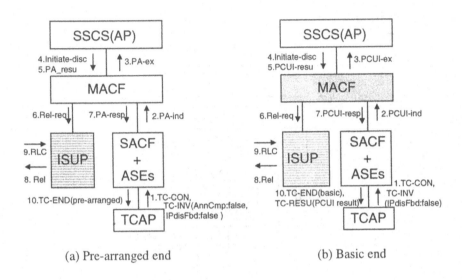

(a) Pre-arranged end   (b) Basic end

Fig. 7. IP initiated disconnection procedures

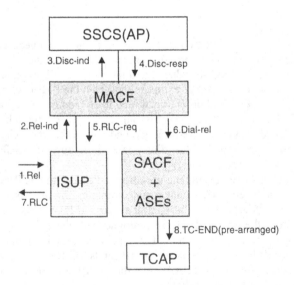

Fig. 8. SSP initiated disconnection procedures

5 Conclusion

We are developing an AIN IP which has an ISUP and CS-1$^+$ INAP for the communication with the SSP and the SCP respectively. At first step, the AIN IP provides specialized resources such as play announcements, collect DTMF, recognize message (i.e., speaker independent from 0 to 9, yes, no, stop), prompt and receive message, message synthesis, and fax process.

In this paper, we proposed an AIN IP architecture which consists of four subsystems distributed into three major processors. Also, we described the telecommunication interface subsystem of IP focusing on INAP such as messages and processing operations and errors. The IP INAP consists of four functional blocks, that is, PMM, SACM, SAOC, and SADC, provides three ASEs and seven operations. To interface with SCP INAP, more concrete decisions have to be made with respect to the AC name, timer values, and operation parameter tag values. In addition, research into protocol conformance testing will be carried out with respect to the already designed INAP functions.

References

[1] James J. Garrahan, Peter A. Russo, Kenichi Kitami, and Reberto Kung, "Intelligent Network Overview," IEEE Communications Magazine pp. 30-36, Mar. 1993.

[2] T.Magedanz, R.Popescu, "Towards Intelligence on Demand - On the Impacts of Intelligent Agents on IN", ICIN'96, Nov. 1996, pp. 30-34.

[3] Alix Leconte et al, "IN value added - the role of IP/SN in IN and Multimedia Applications", ISS'95, April 1995, pp. 206-210.

[4] Go-Bong Choi, Chimoon Han, and Chu-Hwan Yim, "Call Control of Intelligent Peripheral Based on Neural Networks," ISS'95, Vol. 1, pp. 288-292, Apr. 1995.

[5] ITU-T IN CS-1 Recommendations, Q.1218, Geneva, Apr. 1995.

[6] ITU-T IN CS-2 Recommendations, Q.1228, Miyazaki, Jan. 1996.

[7] Korea Telecom, Korea INAP Technical Standards, Aug. 1996.

IN Evolution to Support Mobility in DECT Access Networks

S. Biacchi, G. Ferrari, R. Gobbi

ITALTEL - Central Research Labs
20019 Castelletto di Settimo Milanese (MI), Italy
roberta.gobbi@italtel.it

Abstract. This paper presents the solution given by the AC013 EXODUS European project to the problem of full terminal and personal mobility provision. First by means of IN, the EXODUS projects tracks a possible way to get the foreseen world of UMTS (Universal Mobile Telecommunication System), from a fixed network perspective. A brief introduction to the actual trends of the telecommunication world is given, then the DECT radio interface which is the one chosen by EXODUS is described. The description of the EXODUS network implementation is fully detailed, and the foreseen evolution towards UMTS presented.

1 Introduction

In our society, user mobility has become an essential feature. Nowadays, the portable phone is still the only means of mobile communication. But, as in fixed networks, computers are playing an ever increasing role and it is merely a question of time before they are capable of having these functionality's and more.

The support of mobile services, in particular, in any kind of network is getting more and more important. Users are demanding B-ISDN services to be accessible anywhere and at anytime, and, moreover, they are demanding the introduction of flexible low-cost solutions.

ETSI is in the process of defining the Universal Mobile Telecommunications Service (UMTS) [1],[2]. This is a concept for a world wide standard of third generation mobile telecommunications, enabling mobile voice and data communications in almost any location. UMTS is intended to be universal in the sense of universal coverage, which will include a wide range of services and types of access.

In the framework of the European research programme ACTS, a number of projects are dealing with service aspects and mobility support. The ACTS project AC013 EXODUS is implementing a smooth transition solution towards UMTS and investigating full mobility support in a high speed core network. In EXODUS full mobility means the provision of mobile services in both wired and wireless environments [3].

This paper describes the EXODUS solution for full mobility support of advanced services in the core network, and the interworking towards an existing second generation access system as DECT.

2 The Digital Enhanced Cordless Telecommunication (DECT)

The Digital Enhanced Cordless Telecommunications (DECT) standard was developed by the ETSI Sub-Technical Committee (STC) Radio Equipment and Systems 03 (RES-03) and is specified in ETS 300 175. DECT supports both telephony (voice) and high speed data applications and these can be easily, cost effectively and simultaneously supported in the same terminal. This makes DECT one of the most powerful and important standards for communications. The range of applications for DECT are almost limitless and include Wireless PBX, ISDN, GSM, Telepoint, Wireless Local Loop, LAN, Modem, Fax, E-mail, WWW, X.25 and Frame Relay. DECT is able to support such a wide variety of applications due to the enormous flexibility of its advanced protocol.

A DECT access network is comprised of a Common Control Fixed Part (CCFP) that controls a number of base-stations called Radio Fixed Parts (RFPs). The handset or Portable Part (PP) may also be a PC plug-in card or other device as could be the case in cordless data communications. The DECT access network can be easily attached to a core network, such as PSTN or ISDN as it provides a standardised means of wireless interfacing (interworking) to different networks, including both voice and data networks. DECT uses a TDMA structure which allows up to 12 simultaneous duplex voice/data calls per DECT transceiver. Due to its advanced radio protocol DECT is able to offer widely varying bandwidths by combining multiple bearers into a single channel. This is very useful in data transmission environments for which DECT can support net data throughput rates of n x 24 kbit/s, up to a maximum of 552 kbit/s. As DECT was designed to operate in Residential, Business and Public environments, it also supports full authentication and encryption, thus ensuring that it is a suitable medium for confidential information.

One of the great strengths of DECT is the level of standardisation which has been developed to allow interoperable access to multiple voice and data services via multiple networks. This standardisation is based around the DECT base standards, interworking profiles and full test specifications.

DECT is able to provide in a standardised way, not only voice i.e. a constant bit rate, constant delay (isochronous) service, but an extensive range of circuit switched and packet switched data. The capability to transfer packet data over DECT Common Air Interface drastically boosts its efficiency differentiating it from most of current generation radio access technologies.

The standards for DECT were primarily elaborated for wireless telecommunications (including speech and data) in the private and business environment. More recently, the public environment including radio in the local loop and public mobile networks has gained increasing importance. The Generic Access Profile (GAP) ensures full interoperability between radio installations of different manufactures and the arrival of the standardised CTM Access Profile (CAP) is imminent. The evolution towards wireless multimedia services with bit rates of up to 552 kbit/s within pico-cells is foreseen.

DECT is appropriate for mobile communications within indoor environments and for a limited range of outdoor applications. In outdoor environments multipath interference is a serious problem causing a rapid degradation of service quality with increasing cell radius. For digital cordless systems which are not equipped with equalisers, multipath interference cannot be compensated and as a consequence outdoor deployment is limited to Radio in the Local Loop and mobile public service provision for pedestrians or low speed (30 Km/h) vehicles. DECT uses a dynamic radio channel selection scheme and supports traffic of up to 10000 E/km2 for a cell radius of about 10 meters. The traffic capacity decreases with increasing cell spacing, and at a cell radius of about 100 meters the maximum traffic capacity of DECT is below that of GSM.

3　EXODUS Services

The EXODUS service is the Cordless Terminal Mobility (CTM) service, an ongoing standard within ETSI [1],[3]: the CTM service is based on DECT technology and allows users with cordless terminals to be reached and to make calls (voice and data in the EXODUS network) in the public and private environment, where radio coverage is provided. When the user is outside the radio coverage area or the cordless is switched off the 'Back-up Number' feature enables the user to redirect the incoming calls to a number chosen by the user himself.

Moreover, the service provided by the EXODUS network supports enhanced features with respect to those defined by the standard CTM service as service profile interrogation/modification, roaming between different environments (public, business and residential), different types of access (both fixed and mobile), intra-switch handover, UPT (Universal Personal Telecommunications) call from fixed and wireless terminal, data services to wireless users and wide area business services.

In particular, the provision of data service over DECT to public user has been realised following the approach of transparent provision over the In DECT channel: end-to-end error protection protocol has been developed allowing full compatibility with the installed DECT access system for voice provision.

The EXODUS service addresses enhancements of IN-CS1 capabilities for the support of terminal mobility. Integration of mobility management functionality in IN will make mobility provision more independent from the backbone network.

Moreover EXODUS approach represents the first attempt to design a mobile network as a system that allows true integration of mobile and fixed communications into a single telecommunications infrastructure. In this view EXODUS anticipates the concept of design UMTS as an integral part of B-ISDN.

In the following, a phased model is applied to describe the evolution scenario from CTM+IN towards UMTS [2]. Each step incorporates new functionality that make the system closer to the target UMTS system.

4 The IN Challenge

The aim is to implement and provide mobility in a way as much as possible independent from the air interface.

It will be shown that this service combination supports full personal mobility in fixed networks, as well as full terminal and personal mobility in all DECT networks connected to the service. It has also to be noted that, to provide this service, not only "call related" messages have been used: some requirements have been identified both on IN side and on DECT system and the analysis of the requirements suggests to concentrate the efforts on CS-2 approach, considering also "non call related" messages to support terminal mobility. In reality IN already supports the personal mobility (e.g. in the Universal Personal Telecommunication service) and its architecture is naturally prepared to locate ubiquitous users and to provide them with subscribed services.

On the other hand, in the mobile communication field, the use of mobile terminals and automatic registration procedures facilitates the user's life, avoiding manual procedures to communicate the user's location to the network. Among the existing mobile networks and radio accesses, the cordless air interface DECT has been considered, being a European standard and offering a very high capacity and large range of services support, as e.g. all the ISDN services.

This combination of UPT/IN features and DECT access would link the cordless islands together into a very dynamic and flexible environment, providing to users a complete mobility among different DECT islands, but also between cordless and fixed accesses by means of UPT procedures, offering the services available within IN (e.g. variable routing, messages storing, call record charging, etc.). This service could have a large commercial potentiality, due to user friendliness and due to the availability of small portable handsets.

Finally, the Terminal Automatic Registration process allows to update the network location information as the users move. This process makes end users known to the network and enables incoming call delivery.

When the user is not reachable (DECT terminal switched off or outside the coverage area) the incoming calls are routed to a fixed access, towards a default back up number. This number has to be specified by the user at the subscription time and can be updated via a user procedure corresponding to the UPT registration for incoming calls. Furthermore, the final routing decision can be made taking into account different features subscribed, such as Variable Routing or Call Screening verifications. In case of user not reachable (e.g. terminal switched off) or "not disturb" condition the call could be routed towards a centralised answering system.

Different network architectures may be envisaged to provide full mobility to the users, spanning from dedicated overlay solutions, as in traditional cellular systems, till a fixed network - full integrated solution exploiting the IN capabilities of this network. The IN plays a fundamental role in the deployment of these services. In particular, it allows the provision of these services exploiting the fixed network infrastructures and the integration among the mobility and the PSTN/IN services. The latter approach is presently largely pursued, as it gives the operator the chance to tailor the mobility services offer packages to the needs of customers with cost suited for a mass market (particularly price sensitive).

The actual implementation of these challenging services may be done in several ways and with different degrees of comprehensiveness, depending on the systems used, the technology advancements and the requirements of the targeted market. For example, personal mobility in the wired access network doesn't appear particularly attractive for mass market and will not be referenced to in the following paragraphs. Instead, in this paper a wireless access is considered with radio limited coverage to provide local mobility services in particular zones such as metropolitan/urban areas, highways fuel stations, tourist villages,...

The main objective is the expansion of the current IN architecture including the functional entities needed for the support of the mobility procedures of UMTS; such a complex set of requirements is not yet stable and does not allow the definition of a standard in the short term time frame; nevertheless the IN/UMTS architecture could be used as a vision to be pursued when defining the additional capabilities for the complementation of the Core INAP protocol.

Several network operators are finding the best suited solution at national level. Along this line, activities have been undertaken in Italy to complement the deployed IN with functionality required for the mobility services.Two major concepts were selected as developments guidelines: the first one being a flexible design approach able to favour the evolution towards a standard network architecture as soon as available; the second being the reuse of existing components made available by previous deployments like IN, GSM and ISDN as well as of consolidated interfaces like MAP and INAP national version.

CTM allows the possibility to establish a person to person calls independent of the location, the terminal used and the network access technology (wired and wireless); moreover they involve the offering of a range of telecommunication services able to be flexibly tailored and packaged to meet the needs of individual customers.

4.1 EXODUS Functional Model

Hereafter, the approach followed by the EXODUS Project is described, covering the network aspects and illustrating the solution for an implementation based on the existing Italian IN and the enhancements of Italtel UT100 exchange.

The functional architecture proposed is shown in Figure 1. This architecture is in line with the one proposed in the European standardisation bodies (ETSI CTM Project, ETSI NA6) [5].

• Some important issues can be pointed out: the SCF and the SDF functions are split in two functions each, one dealing with Service Logic and the other dealing with Mobility Management. This allows a more consistent approach to the objective of mobility provision as an IN service: in this way all the mobility related procedures can be managed in an efficient way.

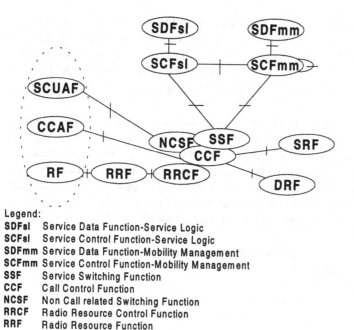

Legend:
SDFsl Service Data Function-Service Logic
SCFsl Service Control Function-Service Logic
SDFmm Service Data Function-Mobility Management
SCFmm Service Control Function-Mobility Management
SSF Service Switching Function
CCF Call Control Function
NCSF Non Call related Switching Function
RRCF Radio Resource Control Function
RRF Radio Resource Function
SRF Specialized Resource Function
DRF Data Resource Function
SCUAF Service Control User Agent Function
CCAF Call Control Agent Function
RF Radio Function

Fig. 1. EXODUS Phase I Functional Architecture

These functionality are fully described in the following:

- **Service Control Function for service logic (SCFsl)**: it is the function that commands call control functions in the processing of IN service requests. The SCFsl may interact with other functional entities to access additional logic, or to obtain information (service or user data) required to process a call/service logic instance. While the SCFmm located in the LEX (Local EXchange) interfaces and interacts with SSF/CCF, SRF and SDFsl functional entities, the SCFsl included in the HDB interfaces only with its correspondent SCFsl inside the LEX.

- **Service Data Function for service logic (SDFsl)**: the SDFsl contains customer and network data for real time access by the SCF in the execution of IN service request. It interfaces with the SCFsl.

- **Service Control Function for mobility management (SCFmm)**: SCFmm is the function that controls mobility management procedures such as location registration and authentication parameters computation. It is organised in hierarchical levels with a centralised (HDB) level and distributed (LEX) level. It interfaces and interacts with other SCFmm for data exchange; it interfaces and interacts with SCFsl to obtain information required to process a call/service logic instance. It interfaces with the SDFsl and SCFsl.

- **Service Data Function for mobility management (SDFmm):** it contains location data (i.e. serving VDB numbers...) and security data (e.g. Authentication keys). It provides them on demand to its correspondent SDFmm located in an interrogating LEX on mobile terminated call. It also receives and stores data related to user location that are used and accessed during incoming call routing. Thus it interfaces with SCFmm.

Another note is the presence of three functional entities (RF, RRF, RRCF) totally devoted to the radio resource management on one side, and the presence of the SRF functional entity that allows the user interaction with the IN services on the other side.

In this way the signalling architecture is based on the Intelligent Network Application Part (INAP) to support personal mobility and supplementary services and an adaptation of the Mobile Application Part (MAP) to support terminal mobility. In this way support of both personal and terminal mobility as IN services can be achieved.

The particular services/features that the project has implemented are:
- incoming and outgoing call;
- location handling;
- handover handling;
- location update;
- location registration/de-registration;
- personal number;
- terminal mobility (whenever radio coverage is provided);
- terminal authentication;
- service profile interrogation/modification;
- roaming between different environments (public, business and residential);
- different types of access (both fixed and mobile);
- Centralised Voice Mail and some supplementary services;
- Wide area Centrex;
- data services.

4.2 EXODUS Network Architecture

The following Figure 2 shows the proposed network architecture. In the figure the physical interfaces are pointed out and they all are explicated in the subsequent Table 1.

A brief description of the network elements is here provided:
- Local Exchange: it is the N-ISDN based local switch. It provides the functionality of non-call related messages handling, Service Switching Point to access the intelligent network, Mobility Management, Call Control and Common Call Control Fixed Part to control the radio fixed part in the access network.
- Radio Fixed Part: it provides the functions of the radio base station for DECT access. Namely some of these functions are: radio frame management, interface to the cluster control fixed part inside local exchange, data support;

The approach followed in line with the D profile is to allow maximum flexibility, to the user and to the developer, in the choice of U-plane protocol used for error correction. Anyway the possible choice PPP (Point to Point Protocol) protocol and possible enhancements to it have been considered. In figure 3 the overall architecture necessary to provide the data service is reported.

It must be underlined the necessity for the LEX to communicate with DECT terminals and with the Application Gateway.

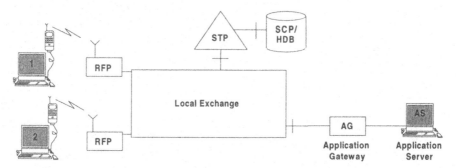

Fig. 3. Network Architecture for data service provisioning

It can be observed that the functional entities involved in the service are the SSF, CCF and RRCF functions. This is because the switch must be able to recognises the radio data call, to manage it and route it to the AG.

5   EXODUS Trials

The architecture and services described in the previous section will be field trial on May '97. The objective of the User Trial is to allow real users to apply EXODUS applications and services in their daily work, and to utilise EXODUS applications and services at locations of their choice. From a requirement point of view, this activity is very distinct from the system demonstration to real users, because user trials require products in a development stage that are near to market introduction (in terms of offered service and QoS). Therefore, only a selection of the applications and services that are developed and demonstrated in EXODUS is suitable for trials with real users.

The physical network infrastructure of the experiment (shown in figure 4) spans two islands, based on two switching centres located in Italy: one in Turin (at CSELT premise) and one in Milan (at Italtel premises). The two sites are interconnected via PSTN.

A number of Radio Fixed Parts directly connected to the exchanges will provide the appropriate radio coverage to the two sites. Fixed access will be added to allow emulation of a complete domestic environment (including both fixed and cordless access), and to demonstrate call completion services and UPT calls.

The mobility management will be performed by a single server called HDB-SCP (Home Data Base - Service Control Point). The connection between the Turin exchange and the HDB/SCP is supported by dedicated digital links.

A Data Server (Application Server & Application Gateway) manages all the data calls and also support access to the Internet.

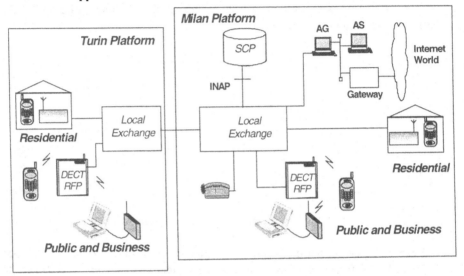

Fig. 4. Physical infrastructure of the network

The experiments will include one "User Trial", two "Field Tests" and at least one "Demonstration to Public". All these experiments will be strongly related to each other, in the sense that they will be based on the same network infrastructure.

5.1 User Trial

A large number of real users will use the wireless DECT access in place of their usual POTS. They will be able to use the service both in their office (indoor environment) and outdoor. The results of the user trial will be deduced by a number of questionnaires filled in by each user every week. The questionnaires will be available in an electronic format. This allows an easier and faster distribution, filling in, and elaboration.

The switch allows the collection of information regarding the traffic generated by the user. This kind of structured information will be collected in order to integrate the subjective information provided by the users. It's important to stress the fact that the objective information available for the User Trial is essentially that provided by the counters implemented in the switch; and this data is not enough to evaluate all phases of the call (e.g., radio access phase, information transfer phase).

5.2 An Implementation Example

In this section an example of the effective implementation of an EXODUS service is described in order to demonstrate how the chosen architecture works implemented.

In particular, the example shows how the basic service of the incoming call is realised. This service has been chosen because it involves all the significant network elements devoted to the mobility management in the service provision.

The incoming call procedure described here assumes a fixed calling user and shows a successful result (the called user is located in the Location Area where is currently registered and the call is answered).

Figure 5 shows all the network entities and interfaces that constitute the end-to-end model supporting the incoming call procedure.

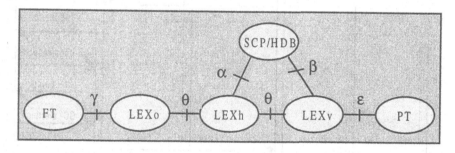

Fig. 5. Incoming call end to end model

The sequence of events that constitute the procedure is listed below and shown in Figure 6, where the messages used are those of the protocols indicated between the different network entities.

The calling user (Fixed Terminal - FT) is attached to a local switch (LEX) not necessarily providing mobility functions, and different both from the home switch (LEXh) and the switch currently visited (LEXv) by the mobile user. On the basis of the called number the call is routed towards the home switch (that is the switch that is able to detect that the called user is a CTM user). Here, the SCP/HDB is invoked to obtain some information about the call treatment. This service node accesses to the user's data contained in the centralised data base and finds the call may be normally routed towards the mobile terminal (Portable Terminal - PT). To do so, it requires to the visited switch to allocate a Roaming number for the CTM user and obtains the routing information as response. These information are forwarded to the home switch so that it can route the call towards the required destination. When the call routing arrives to the visited switch, it starts the interactions with the mobile terminal on the basis of the Routing number contained in the connection message. Assuming the paging phase is successfully executed, a connection between the switch and the mobile terminal is created, so that the call may correctly take place.

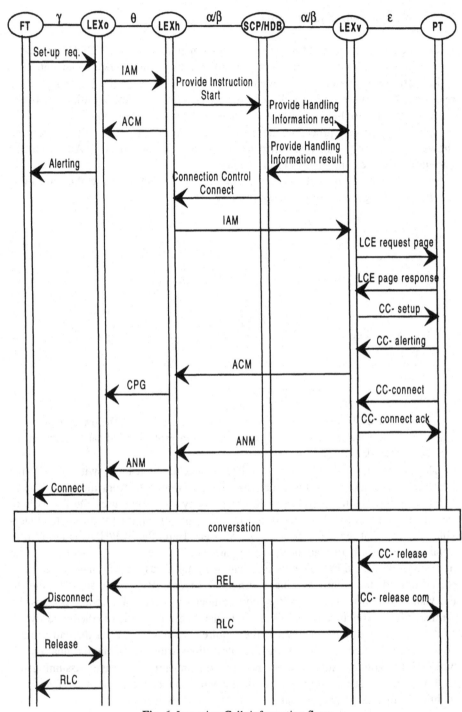

Fig. 6. Incoming Call: information flows

6   Evolution Scenario Towards UMTS

In order to succeed in passing from second to third generation systems in a smooth way, a key issue is the re-use of the existing systems. The evolution proposal of the project foresees two phases: the first phase addresses the systems presented in the previous section and represents the starting point of an evolving platform. The second phase considers the provision of a sub-set of UMTS services and features where all the network components, starting from terminals, access network components and core network are enhanced in order to provide personal and terminal mobility in an B-ISDN environment using DECT as wireless access.

The evolution of IN in order to specify UMTS as an integral part of B-ISDN offering mobility functions with the minimum impact on B-ISDN itself, is the key issue addressed in this phase.

The functional architecture [4] proposed in this phase of the project is depicted in Figure 7.

The specific enhancements that bring to this new functional architecture are proposed in order to cater for separation of service and mobility control and integration of terminal functions for both wired and wireless access. The concept of separation is used as a mechanism to address evolution and to derive a flexible functional architecture based on the current IN [6], UMTS [7] and FPLMTS [8] functional models and architectures.

Also another important thing has to be noted in the figure: the interface between the SCFsl and B-SSF functional entities is dotted. The reason is that while this interface is usually used to manage the fixed terminals, EXODUS intends to foresee a future telecommunication world in which all users and terminals are potentially moveable. Thus the architecture proposed by the EXODUS project presents this type of access to the SCP functions: signalling always passes, before, through the SCFmm/SDFmm for mobility management and, then, the SCFsl/SDFsl treat all questions related to service management.

At this stage also enhancements are foreseen on the overall system regarding the service and mobility control functions, the backbone network, the access network functions and the applications as well.

On the control functions side, enhancements are foreseen on the INAP to support personal and terminal mobility addressing IN CS2 and CS3 standard issues and to manage distributed data access.

In particular the following mobility features will be provided to all the users accessing the EXODUS network:
– Location Registration;
– Location De-registration;
– User Registration;
– User De-registration;
– Location Update;
– Service Profile Modification;
– Service Profile Interrogation;
– Handover Handling.

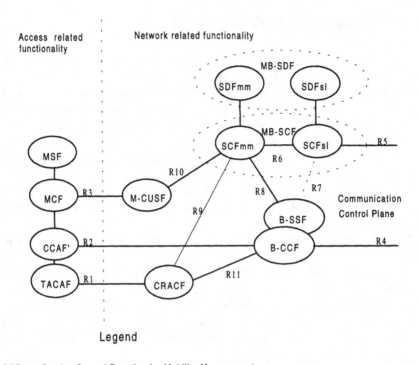

Access related
functionality

Network related functionality

Communication
Control Plane

Legend

SCFmm Service Control Function for Mobility Management
SDFmm Service Data Function for Mobility Management
SCFsl Service Control Function for Service Logic
SDFsl Service Data Function for Service Logic
B-CCF Broadband Call Control Function
B-SSF Broadband Service Switching Function
M-CUSF Mobile Call Unrelated Service Function
CRACF Call Related Access Control Function
MSF Mobile Service Function
MCF Mobile Control Function
CCAF' Enhanced Call Control Agent Function
TACAF Terminal Access Control Agent Function

Fig. 7. EXODUS Phase II Functional Architecture

On the backbone network side, the necessary enhancements over the B-ISDN UNI in order to manage the out of call signalling, the handover and the call handling are foreseen.

From the access network perspective, enhancements on the interworking functions between DECT and B-ISDN in order to manage multi-bearer connections and UMTS location management (especially to manage the separation of user and terminal registration) are implemented.

Finally, multimedia services will be provided to users. In particular services like the following: data base access, multi-media information retrieval, broadband videotelephony and telemonitoring. Two sets of trials targeting highly specialist healthcare application and business users will be run. All of these services are going to be provided in different environments.

Trials are planned in order to implement and evaluate the proposed solution and observe the efficiency of the IN based mobility services.

This just presented is the EXODUS project proposal. As it can be easily seen, it can contribute to many standardisation bodies' work because the work undertaken in the project covers all the aspects of UMTS service provisioning.

Clearly even if this is not the UMTS system fully deployed, the overall project is the road map towards UMTS which is surely the just near next step.

7 Conclusions

The demand of the users for integrated systems providing advanced services and full mobility pushes the definition of UMTS. But changes in networks are evolutionary rather than revolutionary, therefore the concept of a smooth transition path from today's system towards UMTS is selected for the solution presented in this paper.

The technical solution presented is the base for the realisation of a fully integrated network with advanced services enabling terminal and personal mobility.

DECT is currently used for radio in the local loop and for PBX-based DECT systems that are connected to the GSM platform. Public network trials with DECT based mobility have revealed the user interest, benefits and drawbacks of DECT. The main limitation is the radio coverage, which results in restricted service accessibility for the customer.

IN can serve both the fixed and the mobile network and allow the seamless deployment of fixed and mobile services.

Acknowledgements

This work has been carried out within the ACTS project in which the authors are working. However, the views expressed are those of the authors and do not necessarily represent those of the project as a whole. EXODUS is partially funded by the European Commission and by the Swiss Government Office for Education and Research (BBW). The EXODUS consortium consists of Italtel (I), Ascom Tech Ltd (CH), Belgacom, CSELT, GPT Ltd, Kantonsspital Universitatkliniken, Intracom, Laboratoires D'Electronique Philips, National Technical University of Athens (NTUA), Hellenic Telecommunications Organisation (OTE SA), Swiss Telecom PTT R&D, Syndesis, Teltec, Telecom Italia Spa, TMR.

List of acronyms

ACTS	Advanced Communications Technologies and Services
B-ISDN	Broadband Integrated Services Digital Network
MB-SCF	Mobile Broadband Service Control Function
MB-SSF	Mobile Broadband Service Switching Function
CTM	Cordless Terminal Mobility
DECT	Digital Enhanced Cordless Telecommunications
EXODUS	EXperiments On the Deployment of UMTS
FBT	Fixed Broadband Terminal
FPLMTS	Future Public Land Mobile Telecommunication System
IN	Intelligent Network
ITU	International Telecommunication Union
MB-SCP	Mobile Broadband Service Control Point
MB-SSP	Mobile Broadband Service Switching Point
M-SCF	Mobile Service Control Function
UMTS	Universal Mobile Telecommunications System

References

1 Dr. W.H.W. Tuttlebee, "Cordless Communications World-wide", ISBN 3-540-19970-5, Springer Verlag, Aug.96;

2 Special Mobile Group, "Scenarios and considerations for the introduction of the UMTS", ETSI DTR/SMG-050104, May 96;

3 G. Crisp, Changing the emphasis from CTM and Examining the Progress on the standards for CTM, IIR Conference , London, Sept.95;

4 L.Lorenzini, P.Rogl "Mobility in B-ISDN: an IN solution", Interworking '96 proceeding;

5 DTR/NA 61302 (ETSI): Cordless Terminal Mobility (CTM) - IN Architecture and Functionality for the support of CTM (Version 1.11), November 1995.

6 Draft Q.1224 (ITU-T Q.6/11): Distributed Functional Plane for Intelligent Network Capability Set 2, November 1995 version.

7 DTR/NA 61301 (ETSI): IN/UMTS Framework Document (Version 8.1.1). October 1995.

8 Q.FNA (ITU-T Q8/11): Network Functional Model for FPLMTS (Version 1.1.0), 4. - 15. September 1995.

A Mobile Broadband Service Switching Point: A New Network Element for Multimedia and Mobile Services

D. Blaiotta[1,2], L. Faglia[1,3], M. Varisco[1,4], L. Vezzoli[1,5]

[1] ITALTEL Central Research Lab 20019 - Castelletto di Settimo Milanese (MI) - Italy
[2] Donatella.Blaiotta@italtel.it
[3] Lorenzo.Faglia@italtel.it
[4] Monica.Varisco@italtel.it
[5] Lucia.Vezzoli@italtel.it

Abstract. The need to define network architectures and signalling protocols enabling a quick and easy introduction of multimedia services is widely recognized. B-ISDN is the transport infrastructure that has reduced technology constraints in furnishing high-bandwidth transport capabilities. At the same time, users demand for advanced multimedia service to be accessed anywhere and anytime, growing the need of a flexible and low-cost solution to provide mobility on multimedia. In this scenario IN is the forceful mechanism that not only allows a rapid and effective introduction of multimedia services, but also offers the opportunity to add mobility functions to the B-ISDN core network. An architecture of an advanced ATM node, named Mobile Broadband Service Switching Point (MB-SSP), is here investigated. Particular emphasis is given to the three major functional entities allocated within the MB-SSP: the Mobility Handling Function (MHF), the Broadband Call Control Function (B-CCF) and the Broadband Service Switching Function (B-SSF).

1 Introduction

Network operators can adopt for the provisioning to the users of multimedia and mobile services different network architectures. In section 2 alternative scenarios will be briefly described and the reasons for the preference of the IN/B-ISDN architecture are given: it reflects the results of two ongoing Advanced Communications Technologies and Services (ACTS) projects: AC068 INSIGNIA (IN and B-ISDN Signalling Integration on ATM Platforms) and AC013 EXODUS (Experiments On the Deployment of UMTS). Among the numerous topics that are investigated by the above mentioned projects, the characteristics of the Mobile Broadband Service Switching Point network element are here analyzed. The MB-SSP combines the switching capabilities of an ATM node with the functionality that allow to interact with other IN network elements like a Mobile Broadband Service Control Point (MB-SCP) and Broadband Intelligent Peripherals (B-IPs). The network reference architecture is indicated in section 3. The functional model of the MB-SSP is extensively detailed in sections 4 and 5: in section 4, particular emphasis is given to the innovative object model (IN Switching State Model or simply IN-SSM) that realizes the proposed integration between IN and B-ISDN; in section 5, the model of the functions for the handling of mobility is described. The mapping of the functional model of the MB-SSP in the blocks used for the implementation design is contained in section 6. Section 7 summarizes some results both in terms of output to standardization bodies and performance evaluation.

2 Architecture Alternatives

As the market of TLC services is evolving in a difficult predictable direction, a key role is played by the network architecture that must be flexible to cope with service requirements that become relevant after the design of the network itself has been planned. It is important to save investments in the existing equipment and therefore to find a path for a smooth evolution of the network. In the following, some possible scenarios currently under investigation in the TLC world for the provision of both multimedia and mobility services are indicated together with the context in which the architecture described can be placed.

2.1 Provision of Multimedia Services

The provision of multimedia services requires not only the involvement of traditional TLC manufacturers and network operators, but also of new actors (Content Providers, Software houses, Audio/Video Consumer companies) belonging to the wider world of Information Technology (IT).The cost of the terminal equipment and its easy access to the network are crucial aspects to face: if each new offered service requires an expensive terminal and a new interface to the network, the user could consider this solution not interesting both in terms of costs and friendly usage.

Different views are therefore possible: the TLC view is network oriented, as the control and the management of the services are performed inside the network, while the IT view prefers to see the network as a "bit carrier", moving the intelligence at the edge of the backbone. This different philosophy is particularly remarkable when applied to the control architecture that allocates the network resources. In the standardization bodies, the traditional TLC view is pushed by ITU-T when defining the B-ISDN Control Plane. ITU-T SG XI has specified a standardization path for the B-ISDN signalling system which goes through three separate phases, or Capability Sets (CS). CS1 offers the capability of controlling point-to-point monoconnection basic calls. A single CS1 call is not satisfactory for the provisioning of multimedia services, that usually require multiconnection and multipoint call configurations. Only with CS2 and CS3 the set of advanced required capabilities will be available. At the moment only CS1 protocols are commercially available. The "full signalling" solution is not the only possibility because also the Intelligent Network infrastructure can play a significant role in handling multimedia services: the IN functional architecture is able to discriminate the basic functions provided by the telecommunication network (resource allocation and transport of bearer services) from the managing of the features of specific telecommunication services. Integrated IN and B-ISDN architectures are objective of B-ISDN and IN CS3, so this solution is a long term approach. Another standardization body, named Digital Audio-Visual Council (DAVIC), has proposed a compromise between the TLC and IT views. Given a network infrastructure with B-ISDN CS1 signalling capabilities, all the additional signalling requirements of multimedia services are covered introducing a session control level (and related session protocols like Digital Storage Media - Command and Control User-Network) between the end systems (Set Top Unit, Servers) and a network element controlling the resources (Server Call/Connection entity).

Given these three scenarios ("full signalling", integrated IN/B-ISDN and DAVIC scenario), the preference within the EXODUS and INSIGNIA projects has been given to the integrated IN/B-ISDN architecture for the following reasons:

- during the lifetime of a complex service instance, the involved users can dynamically require the addition of new U-plane channels; it is therefore needed the presence of a centralized element (similar to the Server Call Connection entity) that translates the user's requests in the establishment of calls/connections. For the time being, the "full signalling" scenario is not suitable for handling this requirement because the "third party call establishment" capability has not been standardized yet;

- it is a service independent solution: the introduction of new services does not necessarily implies modifications to the terminals and signalling nodes; this is assured by the IN that separates service aspect from the underlying network infrastructure: to add a new service means to realize a new Service Logic;

- the terminals support standard signalling protocols and not ad-hoc solutions for each service; this is not guaranteed by DAVIC approach because the session control added in the terminals to the call/connection control is used only for a restricted set of services .

In other words, the IN/B-ISDN integration seems the most promising approach.

2.2 Provision of Mobile Services

Second generation mobile telecommunication systems, such as GSM and DECT, have been widely deployed throughout Europe and their impact is growing world-wide. GSM offers pan-European roaming in public networks, while DECT systems have been employed in business and domestic environments.

In the research activities and standardisation work, evolutionary third generation mobile systems (UMTS/FPLMTS) supporting multiple radio access systems, are now being considered. These are intended to provide telecommunication services including voice, video and data, to mobile and fixed users via a wireless link, covering a wide range of user sectors (e.g., private, business, residential, etc.) and accommodating a wide range of user equipment. The migration path from existing (second generation) systems towards 3rd generation ones must foresee a smooth transition, through the reuse and the enhancement of existing technologies in order to protect the significant investment in the field. Two approaches are possible: an evolution from mobile network or an evolution from the fixed network. Evolution from mobile networks starts from GSM family networks: GSM 900 and DCS 1800 in Europe and DCS 1900 in USA, with service enhancements from Camel. Evolution from fixed network starts from narrowband public networks: PSTN and N-ISDN, with CTM service based on DECT radio access technology.

The ACTS Exodus project approach described in this paper, is the latter one. It is based on enhancements of the core network capabilities for supporting terminal and personal mobility. In this approach, the Intelligent Network is considered as the enabling technology to implement mobility functions. The application of IN to mobility is a key element in the integration of a mobile access to future intelligent networks such as B-ISDN/IN. In that integration, IN would take care of the user and mobility control with related data, while B-ISDN is responsible for the basic switching and transport mechanism. The key elements of IN that makes it suitable for handling mobility are:

1. IN is able to provide service-independent functions that can be used as "building blocks" to construct a variety of services. This allows easy specification and design of new services.
2. IN provides services independently from the network implementation. This allows to isolate the services from the way the service-independent functions are actually implemented in various physical network infrastructures; in particular, independently from the fact that a network is fixed or mobile. IN allows to have a universal core network regardless of means of access: from a core network point of view a fixed/mobile Interworking Unit and a wireline terminal have the same behaviour.

Following the evolution of IN Capability Sets, different steps for the modelling of mobility functions can be identified. They differ with respect to the integration degree of mobility functions into the IN service logic. The role of intelligence as seen by IN (that is the service logic in the SCPs) increases gradually to embrace the system functions of user and terminal mobility.

3 Network Elements of the Selected Architecture

In Fig. 1 the reference physical architecture is shown. It is based on a switched broadband network integrated with evolved IN functionality. The network elements included in the architecture can be divided into Access Network elements and Core Network elements. The Access Network elements are:

- Terminal Equipment: Fixed Broadband Terminal (FBT) and DECT Mobile Terminals (MT);
- DECT Base Station: it represents the DECT Radio Fixed Part;
- Interworking Unit between DECT and Broadband Network: this unit terminates the DECT protocols and interworks them with ATM protocols.

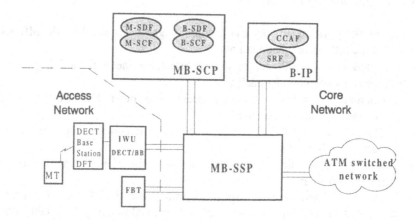

Fig. 1. Physical architecture of the network

From the core network perspective wired and wireless access appear identical.
The Core Network elements are:

- Mobile Broadband Service Control Point (MB-SCP): this element contains the logic necessary in order to support the management of the mobility and to handle multimedia services. To this purpose two different groups of functional entities are identified: *M-SCF/M-SDF* that provides the control of mobility management procedures; *B-SCF/B-SDF* that provides the control of multimedia user initiated services. Routing information provisioning, authentication processing, initiate call establishment and release are some of the functionality offered by the B-SCF. The B-SCF is provided with the knowledge of the overall service characteristics and it instructs the MB-SSP itself about how the service must be handled, also in terms of call configurations. The separation of M-SCF/M-SDF and B-SCF/B-SDF is proposed to allow different type of service logic instance to run at the same time, for the same call. More than one MB-SCP can exist in a international scenario. Each user has a "home network" where all his data are permanently stored, but can access his services from any other "visited network". On the registration of the user into a new network, part of his data are copied from the home SDF to the visited SDF.

- Broadband Intelligent Peripheral (B-IP): this unit is responsible for the provisioning of particular features like the multimedia user to service interaction. It provides the user with the menu navigation capabilities and acts a specialized server for information retrieval. The functional entities allocated in the B-IP are the Specialized Resource Function (SRF) and the Call Control Agent Function (CCAF).

- MB-SSP: it is broadband switch upgraded with IN capabilities enabling the interaction with intelligent nodes (MB-SCP and B-IP).

In the following the description of the MB-SSP is given, not emphasizing the internal structure of the other network elements.

3.1 MB-SSP System Overview

The proposed physical network architecture assigns a central role to the MB-SSP as it provides the following functionality:
- it provides users with access to the broadband network and it performs the necessary switching functionality;
- it contains capabilities to communicate with other MB-SSPs by means of broadband signalling protocols;
- it allows access to the set of IN capabilities. It contains capabilities to detect requests for IN-based services;
- it contains capabilities to communicate with the MB-SCP and to respond to instructions from it;
- it allows the interaction between the terminals and the MB-SCP for handling all the procedures related to the terminal and personal mobility, by establishing a call unrelated logical connection between them. It contains the logic necessary to recognize call unrelated events and to report them to the M-SCF.

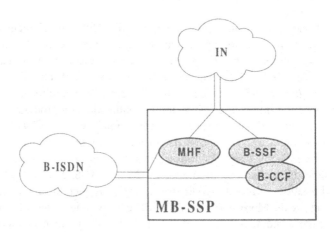

Fig. 2. Functional entities located in the MB-SSP

In Fig. 2 the functional entities located in the MB-SSP are depicted:
- the B-CCF is responsible for handling B-ISDN calls; it comprises call and bearer control components and the capability of establishing/releasing calls and connections according to B-SCF commands. A two level Broadband Basic Call State Model (B-BCSM) has been defined for the representation of signalling application process, in order to detect call and bearer related events relevant for the service logic;
- the B-SSF extends the logic of the B-CCF to be able to dialogue with the B-SCF in the MB-SCP. It is able to coordinate all the triggers coming from the B-BCSMs that compose a single service instance. A novel IN-Switching State Model (IN-SSM) is defined in order to represent the network resources in terms of attributes

and relationships. This view is offered to the B-SCF and it allows the separation of service dependent aspects from network related signalling capabilities;

- the MHF handles the call unrelated events initiated both by the user and by the network in order to perform mobility management procedures. A Mobile Basic Call Unrelated State Model (M-BCUSM) is developed to detect call unrelated events requiring an intelligent processing.

The detailed description of the three functional entities B-CCF, B-SSF and MHF is the object of sections 4 and 5.

4   Call Control and Service Switching Functions

The B-CCF and B-SSF functional entities defined above have relationships with the corresponding CCF and SSF functional entities defined in [1]. The main difference is that they apply to a broadband and not to a narrowband context: this means that a service instance can be composed by more than two parties and more than one call/connection. This implies that the B-SSF has to offer to the Service Logic in the MB-SCP a much more complex view of the set of network resources respect to what requested in narrowband scenarios. This requirement leads to the introduction of an additional control domain (the Session Control Domain) between the Call Control and the Service Control, as detailed in section 4.1. Regarding the B-CCF model the ITU-T reference is [2], where the distinction between call and connection control is introduced. This is reflected here splitting the monolithic Call Control Domain in two domains: Call and Connection Control Domains.

4.1 B-CCF/B-SSF Model

The following four control domains can be distinguished in general:

- *Service Control Domain*: the overall control of a service is carried out by the Broadband Service Control Function.
- *Session Control Domain*: the term session is used to denote an association of calls and connections for the realisation of a single service. This network view of a service is controlled by the IN Service Switching Function (B-SSF).
- *Call Control Domain:* the realisation of a B-ISDN call is managed by a component of B-ISDN switching control which is traditionally named "Call Control" (CC).
- *Connection Control Domain:* the physical switching resources are managed by a component of B-ISDN switching control which is traditionally named "Bearer connection Control" (BC).

The MB-SSP is a service independent network element, so the Service Control Domain is outside its scope. Call Control (CC) and Bearer connection Control (BC) functionality together form the so-called Broadband Call Control Function (B-CCF). In Fig. 3 the above control domains and their mapping towards IN Functional entities are shown. The main enhancement of the proposed B-SSF is the capability of handling sessions. The IN Switching Manager maintains two different views of a service, the overall party view and a specific view of each bearer connection. The party view is used to co-ordinate several bearer connections.

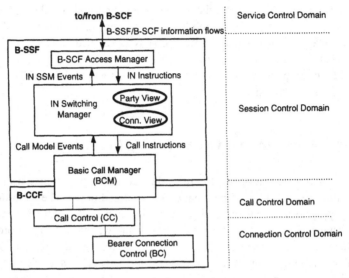

Fig. 3. B-SSF/B-CCF Functional Model

In the following, the functional blocks are explained in more detail:

- the *Basic Call Manager (BCM)* provides an abstraction of a part of a switch that implements basic call and connection control to establish communication paths for users. It detects basic call and connection control events that can lead to the invocation of IN service logic instances or should be reported to active IN service logic instances, and manages B-CCF resources required to support basic call and connection control. The BCM functionality is distributed among B-CCF and B-SSF.

The B-SSF comprises the IN Switching Manager and the B-SCF Access Manager:

- the *IN Switching Manager* keeps information on the status of a IN service by means of the IN Switching State Model that allows to model the connections belonging to a session using the defined objects. It interprets the Call Model Events and translates them into IN SSM events (SSM state changes) to be communicated to the B-SCF. In the reverse direction, IN instructions coming from the B-SCF are translated into instructions to the BCM. Furthermore, the Switching Manager can instruct the B-CCF to start call processing according to instructions from the B-SCF ("SCP-initiated calls" as a particular case of the capability "third party call set up");

- if IN service processing is invoked (either by the B-SSF, on detection of a special condition, or by the B-SCF), communication between the B-SSF and the B-SCF starts. This communication is handled on the B-SSF side by the *B-SCF Access Manager*. The B-SCF Access Manager has the responsibility to locate the required

IN service in the ATM Network and to send messages to and receive messages from the B-SCF.

The domain classification indicated in Fig. 3 leads to a two level approach in the definition of the B-BCSM, as it is possible to distinguish events related to the call from events related to the connection. This means that a service instance is represented in the B-CCF by a set of state machines that can trigger B-SSF in an independent way.

4.2 IN-SSM Model

According to [1], the IN-SSM provides a framework for describing the scope of view and control of SSF/CCF activities offered to an SCF. However, due to the simple topology of IN CS-2 services, the role of SSM is not relevant. In a broadband environment, a generic service can have a complex topology if it is composed by more than a single call/connection. The problem of call/connection correlation has been solved introducing the concept of *session* inside the IN-SSM model. The session view models the co-ordination of the different connections (and therefore different B-BCSMs) involved in providing a single service, as perceived locally in the B-SSF. The IN SM in the B-SSF will handle the requests coming from the B-BCSMs and will correlate them in the context of a session, sending the appropriate messages to the B-SCF. In the other direction, the IN SM will receive the commands coming from the B-SCF and will take the appropriate actions on the B-BCSMs. The IN-SSM is the tool needed to represent the connections, the parties involved in a session and their relationships. In Fig. 4 the graphical representation (using OMT notation) of the new IN-SSM is depicted. The names of the associations have to be read from left to right or from top to bottom.

A *session* is the representation of a complex call configuration, as it is seen by the IN functional entities. Several *parties* can join a session. One of the parties joining a session is the session owner, so at least one party joins a session. During a session, new parties can be added to or existing parties can be removed from a session. *Bearer connections* can be established between the parties of a session. If B-ISDN signalling CS-1 is used, there is a one to one relationship between call and bearer connection. A bearer connection is composed of several *legs*. A leg represents the communication path to a party which is connected to other parties by a bearer connection. The multiplicity of the aggregation relation between leg and bearer connection determines the topology type of the connection. If a bearer connection contains exactly two legs, it is a point-to-point connection. For point-to-multipoint-connection each bearer connection has an association with more than two legs. The attributes of the objects are described in more detail in Table 1. The attribute "Is_Virtual" indicates whether the respective party object is representing a network element (i.e. the SCP, when the attribute values is TRUE) or a true party associated with an end system.

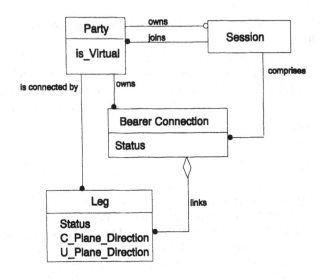

Fig. 4. Object model of the IN-SSM

The "Status" of Bearer Connection indicates in which state is the bearer connection object in a given phase:

- Connection establishment phase   BEING SETUP
- Active phase   SETUP
- Connection release phase   BEING RELEASED

The "Status" of Leg indicates in which state is the bearer connection object in a given phase:

- Leg establishment phase   PENDING
- Route selected phase   DESTINED
- Leg established phase   JOINED
- Calling party initiated release during establishment phase   ABANDONED
- Call refused by the called party   REFUSED

Relationships are obviously present between the status attributes of the bearer connection and the legs: the status of the bearer connection is BEING SETUP, as long as the status attributes of all corresponding legs are not JOINED. Only if all corresponding legs are JOINED, the bearer connection will be SETUP. The IN-SSM view, composed by the SSM objects and their relationship, is communicated from MB-SSP to MB-SCP (or viceversa) through information elements in the Information Flows.

4.3 B-SSF/B-SCF Interface

Information Flows (IFs) between B-SCF and B-SSF either consist of a request/response pair or of a request alone. A relationship between B-SCF and B-SSF is established either as a result of the B-SSF sending a request for instruction to the B-

SCF, or after the B-SCF sends a request for an action on IN-SSM objects. The list of IFs can be found in [14].

4.4 B-CCF/B-SSF Interface

An innovative aspect of this model is that the Service Control Function and the Service Logic do not have a direct view of the B-CCF call states (Detection Points of the B-BCSM). The communication between B-SCF and B-SSF refers to the more abstract notions of sessions, parties, bearer connections and legs as they are represented in the SSM. However, since the Call Control (B-CCF) is defined in terms of a BCSM, some translation is required between the abstract, SSM-related view offered to the B-SCF and the basic view of the B-CCF. This translation is carried out by the IN Switching Manager in the B-SSF functional model. Considering that B-CCF and B-SSF are located in the MB-SSP this means that this interface needs not to be standardised.

5 Mobility Handling Function

The MHF handles the call unrelated events initiated both by the user and by the network in order to perform mobility management procedures. The supported mobility procedures are Location registration/de-registration, User registration/de-registration, Location update, Look Ahead Paging and User profile interrogation and modification. The handling of mobility aspects is treated independently from the access technology: the wired and the wireless access technology, are handled in the core network, that is from a network perspective, in a integrated manner. A mobility management protocol between the terminal and the MB-SCP has been defined. It allows the handling of the call unrelated mobility procedures. This protocol is defined on top of two different "transport" protocol respectively at the UNI between the user and the MB-SSP and at the IN interface between the MB-SSP and the MB-SCP. In order to better understand the description of the MHF mechanisms a short indication of the selected protocols is given in the following.

- at the UNI: for the transport of Mobility Management, the connectionless Bearer Independent (CL-BI) transport mechanism, according to ITU recommendation Q.2932 [6], is proposed;
- at the MB-SSP/MB-SCP interface: the standard CS2 INAP protocol has been enhanced to define a new protocol that supports mobility in a broadband environment; it has been called Mobile-INAP (M-INAP) protocol and it handles the mobility management procedures. M-INAP is based on the ITU work Q.1228 [7].

A Mobile-Basic Call Unrelated State Model (M-BCUSM) is developed to detect call unrelated events requiring an intelligent processing. The M-BCUSM is a high level description of the MHF activities required to establish and maintain an association between users and service processing and to manage invoked operations. It provides

the trigger mechanisms for wired and wireless call unrelated interaction to access IN functionality; it modifies call unrelated interaction processing functions as required to process requests for IN provided service usage under the control of the MB-SCP. Two types of Call Unrelated procedures are distinguished:
1. Terminal Initiated (e.g. User Registration, Location Update);
2. Network Initiated (e.g. Authentication, Network Initiated User Deregistration, Paging).

The transfer of information between the user and a Service Logic is transparent in the procedures initiated by the user. In this case the introduction of new IN mobility services does not imply enhancements of the network. For Network Initiated procedures, the point of attachment of the addressed terminal has to be dynamically determined by the MHF. In particular, in the case of a wireless terminal, a Paging procedure is necessary before it can be determined. An MB-SSP involvement is therefore needed for the network initiated procedures.

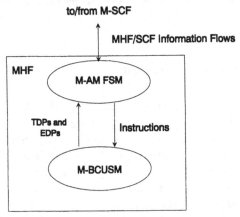

Fig. 5. MHF model

In Fig. 5, the Finite State Machines (FSM) inside the MHF are shown. To model the call unrelated associations established at the UNI, a Mobile-Basic Call Unrelated State Model (M-BCUSM) is proposed. Moreover, a FSM devoted to the communication with external entities (M-SCF), called in the following sections M-SCF Access Manager (M-AM) FSM, that models the relationship between the MHF and the M-SCF during a call unrelated association, is described. The description of M-BCUSM and M-AM FSM can be found in [14].

6 Implementation design

The design of the implementation is decomposed in the design of the functional blocks (section 6.1) and in some details about languages and operating system used for the implementation are reported in [14].

6.1 Functional Blocks

When moving from the functional model to the implementation design it is needed to map the functional entities in the appropriate protocol state machines and application processes. In Fig. 6 the mapping used in the MB-SSP is shown.

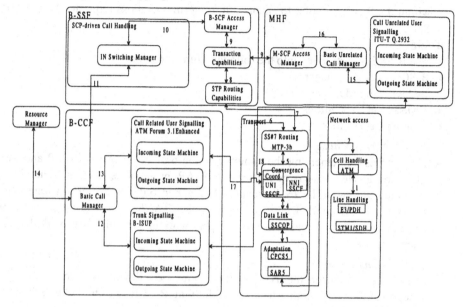

Fig. 6. Functional blocks of MB-SSP

B-CCF
The B-CCF is divided into the following functional blocks:
- Call Related User Signalling;
- Trunk Signalling;
- Basic Call Manager.

Call Related User Signalling
- This block implements access signalling, based on the specifications provided by ATM Forum UNI 3.1 [8] and ITU-T DSS2 Q.2931 [9]. It is divided into two sub-blocks, which respectively behave as *Incoming State Machine* and *Outgoing State Machine*. The terms Incoming and Outgoing refer to the direction (calling to called) of the signalling relation.

Trunk Signalling
- This block implements the network node signalling, a version of the ITU-T B-ISUP Q2761/2/3/4 [10-13]. The same internal subdivision between Incoming and Outgoing State Machine is present.

Basic Call Manager

– The *Incoming* and *Outgoing State Machines* in *User/Trunk Signalling* provide a service to the Application Processes residing in the *Basic Call Manager* function. This functional block also allows interconnection between B-CCF and B-SSF. Additional functions, like network-initiated call, are inserted into this block. The *Basic Call Manager* offers an appropriate interface to the *IN Switching Manager* within the B-SSF in order to enable it to perform operations requested by the MB-SCP.

B-SSF

The B-SSF is partitioned into the following functional blocks:

- B-SCF Access Manager;
- IN Switching Manager;
- Transaction Capabilities;
- Routing Capabilities.

B-SCF Access Manager

– Here, ASEs implementing the communication protocol between B-SSP and MB-SCP are located. In particular, the B-INAP protocol is present. This block communicates with the *IN Switching Manager* in order to perform a transaction towards the MB-SCP. Dialogue between the *B-SCF Access Manager* and the *Transaction Capabilities* ensures proper working of the MB-SSP/MB-SCP communication.

IN Switching Manager

– This block implements the IN-SSM. It performs the following functions:

1. reception of signals coming from the underlying B-CCF or from the underlying *B-SCF Access Manager*;
2. translation of external signals into appropriate action on the SSM (session creation, object creation, object updating, ...);
3. transmission of signals towards the B-CCF or towards the *B-SCF Access Manager*;
4. maintenance of the SSM data structures related to the session and to the session objects.

– The *IN Switching Manager* handles different session simultaneously. The sessions are initiated by triggers coming from the B-CCF. This mechanism can be enhanced in order to enable sessions activated by the MB-SCP. For each session, the *IN Switching Manager* keeps track of the new instances of the session objects and of their state. The mapping between SSM objects and B-CCF detection points is part of the *IN Switching Manager* tasks. The *IN Switching Manager* is given the capability to translate B-INAP operations reported by the *B-SCF Access Manager* into commands directed towards the B-CCF. In particular, it exploits the B-CCF interface in order to ask for call routing, call setup/tear down, detection point arm/release. Communication with the *B-SCF Access Manager* takes place in order to initiate transactions and to perform MB-SCP commands.

Transaction Capabilities

– It includes the TCAP protocol for the support of MB-SSP/MB-SCP transactions. This block communicates with the *B-SCF Access Manager*, from which it receives/to which it sends TCAP primitives containing B-INAP messages, and with the *STP Routing Capabilities*, to which it delivers/from which it receives the message going to/coming from the MB-SCP.

Routing Capabilities
- This block is useful when the MB-SSP and the MB-SCP are not directly connected and a (network of) STP(s) is present. In this case, routing within the network is performed. The connectionless mode of operation is supported.

MHF

The MHF is divided into the following functional blocks:
- Call Unrelated User Signalling;
- M–SCF Access Manager;
- Basic Call Unrelated Manager;

Call Unrelated User Signalling
- This block implements CL-BI part of the ITU-T Q.2932 [6] signalling. It is divided into two sub-blocks, which respectively behave as *Incoming State Machine* and *Outgoing State Machine*.

M-SCF Access Manager
- Here, ASEs implementing the communication protocol between MB-SSP and MB-SCP related to mobility management, are located. In particular, the M-INAP protocol is present. This block communicates with the *Basic Call Unrelated Manager* in order to perform a transaction towards the MB-SCP. Dialogue between the *M-SCF Access Manager* and the *Transaction Capabilities* ensures proper working of the MB-SSP/MB-SCP communication.

Basic Call Unrelated Manager
- This functionality communicates with *M-SCF Access Manager* and with the *Call Unrelated User Signalling* and provides the necessary capabilities for the interworking of the protocols.

Finally, the *Resource Manager* does not belong to any Functional Entity: it handles the Connection Termination Point (CTP) and Cross-connection (XC) resources of the switching system.

7 Results

The results can be evaluated both as an impact on standardadization bodies and as a trial evaluation. The functional model has been presented in standardization bodies [3, 4, 5], where the integration of IN and B-ISDN is an item that is starting to be considered, even if only in a long term perspective. Interest for the proposed model has been expressed by the IN experts. The proposed architecture must not only be consistent from the functional point of view, but it must offer an acceptable behavior from the performance point of view. The work in this area is split in theoretical investigation and measures derived from the trials. Regarding the modeling, particular attention has been paid to the evaluation of the impact of the innovative concepts, such as the support of the MB-SCP initiated call capability and the concept of session.

Some performance issues of particular relevance in an IN based scenario have been outlined. In particular, the need of a congestion control mechanism able to avoid service degradation which can occur due to overload situations has been stressed. Another issue which has been outlined is the system scalability. By this expression it is indicated the system capability to support increases in number of users, network nodes, network coverage and services. A definition of the traffic scenario has been derived, characterizing the traffic on the basis of the services description. The fundamental parameters for the service characterization have been singled out, and preliminary hypotheses on their values have been done. The ongoing trials are producing a feedback to the theoretical results according to the defined set of measures which should provide a clear idea of the efficiency of the IN signalling network. An example of a service scenario adopting this network architecture can be found in [14].

8   Conclusion

The advanced MB-SSP network element considered in this paper realizes a set of new functionality associating to the Call Control Function for Broadband signalling two powerful functional entities used for the dialogue towards IN:
- an evolved B-SSF with an object oriented approach for the communication with the MB-SCP;
- an MHF for managing the mobility procedures.

It has been shown how the model is mapped in functional blocks. The MB-SSP is able to face the requirements of a large variety of services, both in multimedia and in mobile environment.

List of acronyms

ACTS	Advanced Communications Technologies and Services
B-BCSM	Broadband Basic Call State Model
B-CCF	Broadband Call Control Function
B-IP	Broadband Intelligent Peripherals
B-ISDN	Broadband Integrated Services Digital Network
B-SCF	Broadband Service Control Function
B-SSF	Broadband Service Switching Function
CTM	Cordless Terminal Mobility
CTP	Connection Termination Point
DAVIC	Digital Audio-VIsual Council
DCS	Digital Cellular System
DECT	Digital Enhanced Cordless Telecommunications
EXODUS	EXperiments On the Deployment of UmtS
FBT	Fixed Broadband Terminal
FPLMTS	Future Public Land Mobile Telecommunication System

GSM	Global System for Mobile
IN	Intelligent Network
INSIGNIA	IN and B-ISDN SIGNnalling Integration on ATM platforms
IN-SSM	Intelligent Network Switching State Model
IPC	Inter Process Communication
ITU	International Telecommunication Union
M-BCUSM	Mobile Basic Call Unrelated State Model
MB-SCP	Mobile Broadband Service Control Point
MB-SSP	Mobile Broadband Service Switching Point
MHF	Mobility Handling Function
M-SCF	Mobile Service Control Function
MT	Mobile Terminal
OMT	Object Modeling Technique
PIA	Point In Associations
UMTS	Universal Mobile Telecommunications System

References

1. ITU-T SG11 Rec. Q.1224 *"Distributed Functional Plane for Intelligent Network Capability Set-2"* - Geneva, January 1997
2. ITU-T SG11 COM11-R35 *"Broadband Capability Set 2 Signalling Requirements"*
3. Delayed contribution D.5 to ITU-T SG11, Geneva 13-31 January 1997 "Relationships of functional entities to the SCF in the Unified Functional Model"
4. Delayed contribution D.4 to ITU-T SG11, Geneva 13-31 January 1997 "Proposal for an IN Switching State Model in an integrated IN/B-ISDN scenario"
5. Temporary Document TD 47 to ITU-T SG11 WP3, New York 7-11 April 1997 " Presentation of the EXODUS project"
6. ITU-T SG11 Rec. Q2932.1 *"Digital Subscriber Signaling System No. 2 (DSS2) Generic functional protocol: Core functions"* - Geneva, July 1996
7. ITU-T SG11 Rec. Q.1228 *"Interface Recommendation for Intelligent Network CS 2"* - Geneva, January 1997
8. ATM Forum - User-Network Interface (UNI) Specification Version 3.1 - September 1994
9. ITU-T SG11 Rec. Q.2931 *"Digital Subscriber Signaling System No. 2 (DSS2) - User network interface (UNI) layer 3 specification for basic call/connection control"* - Geneva, February 1995
10. ITU-T SG11 Rec. Q.2761 *"Functional description of the B-ISDN user part (B-ISUP) of Signaling System No. 7"* - Geneva, February 1995
11. ITU-T SG11 Rec. Q.2762 - *"General functions of messages and signals of the B-ISDN user part (B-ISUP) of Signaling System No. 7"* - Geneva, February 1995
12. ITU-T SG11 Rec. Q.2763 - *"Signaling System No. 7 B-ISDN user part (B-ISUP) of Signaling System No. 7 - Formats and codes"* - Geneva, February 1995
13. ITU-T SG11 Rec. Q.2764 - *"Signaling System No. 7 B-ISDN user part (B-ISUP) of Signaling System No. 7 - Basic call procedures"* - Geneva, February 1995
14. D. Blaiotta, L. Faglia, M. Varisco, L. Vezzoli -*"A Mobile Broadband Service Switching Point: a new functional architecture for multimedia services"* - AIN'97 2nd International Workshop on Advanced Intelligent Networks - Cesena - July 1997

Author Index

Springer
and the
environment

At Springer we firmly believe that an international science publisher has a special obligation to the environment, and our corporate policies consistently reflect this conviction.

We also expect our business partners – paper mills, printers, packaging manufacturers, etc. – to commit themselves to using materials and production processes that do not harm the environment. The paper in this book is made from low- or no-chlorine pulp and is acid free, in conformance with international standards for paper permanency.

 Springer

Lecture Notes in Computer Science

For information about Vols. 1–1308

please contact your bookseller or Springer-Verlag